Media Seas and the Blue Humanities

This book contributes to the emerging field of the blue humanities by developing a critical media perspective on the technical, social, cultural, and historical mediation of knowledge, perception, and experience of the oceans. With contributions from leading scholars from across a diverse range of fields, this volume focuses on the environmental challenges experienced in the northern reaches of the North Atlantic Ocean. The authors define these areas as the High North Atlantic, characterized by significant temperature fluctuations and considerable geopolitical significance, and noted for experiencing many of the most severe impacts of climate change.

Case studies consider the environmental impacts of offshore oil, fish farming, deep-sea mining, and other extractive industries as a result of operations conducted by neighboring countries such as Norway, Denmark, the UK, Iceland, Greenland, the USA, and Canada. A key focus is on technical mediation: the interactions between technologies, humans, and marine ecosystems, and by deploying a critical media perspective throughout, the book explores the construction of aquatic cultures and the ways in which maritime technologies organize ways of seeing, sensing, and comprehending the changing natures of the oceanic world.

As an important addition to the blue humanities and environmental humanities in general, this interdisciplinary volume will appeal to scholars and students of media and communication studies, human geography, anthropology, environmental studies, cultural and technological history, visual studies, architecture, art history, and literature.

Jørgen Alnæs is Associate Professor at the Department of Journalism and Media Studies, Oslo Metropolitan University, Norway.

Synne Tollerud Bull is Professor and Pro-Dean of Research at School of Art, Design, and Media, Kristiania University of Applied Sciences, Norway.

Liv Hausken is Professor of Media Studies at the University of Oslo (UiO) Department of Media and Communication, Norway.

Susanne Østby Sæther is Senior Curator of Photography and New Media at Henie Onstad Art Center, Norway.

Routledge Environmental Humanities

Series editors: Scott Slovic (University of Idaho, USA), Joni Adamson (Arizona State University, USA) and Yuki Masami (Aoyama Gakuin University, Japan)

The Routledge Environmental Humanities series is an original and inspiring venture recognising that today's world agricultural and water crises, ocean pollution and resource depletion, global warming from greenhouse gases, urban sprawl, overpopulation, food insecurity and environmental justice are all crises of culture.

The reality of understanding and finding adaptive solutions to our present and future environmental challenges has shifted the epicenter of environmental studies away from an exclusively scientific and technological framework to one that depends on the human-focused disciplines and ideas of the humanities and allied social sciences.

We thus welcome book proposals from all humanities and social sciences disciplines for an inclusive and interdisciplinary series. We favour manuscripts aimed at an international readership and written in a lively and accessible style. The readership comprises scholars and students from the humanities and social sciences and thoughtful readers concerned about the human dimensions of environmental change.

For more information about this series, please visit: www.routledge.com/Routledge-Environmental-Humanities/book-series/REH

Media Seas and the Blue Humanities

Studies of the High North Atlantic

Edited by
Jørgen Alnæs, Synne Tollerud Bull, Liv Hausken and Susanne Østby Sæther

Routledge
Taylor & Francis Group
NEW YORK AND LONDON

First published 2026
by Routledge
605 Third Avenue, New York, NY 10158

and by Routledge
4 Park Square, Milton Park, Abingdon, Oxon OX14 4RN

Routledge is an imprint of the Taylor & Francis Group, an informa business

Library of Congress Cataloging-in-Publication Data
A catalog record for this title has been requested

ISBN: 978-1-032-76512-9 (hbk)
ISBN: 978-1-032-76515-0 (pbk)
ISBN: 978-1-003-47880-5 (ebk)

DOI: 10.4324/9781003478805

Typeset in Times New Roman
by Taylor & Francis Books

Contents

Figures

Acknowledgments

This book emerges from the international research collaboratory Media Seas of the High North Atlantic, convened and led by the editors of this book and supported by the Oslo School of Environmental Humanities (OSEH) at the Faculty of Humanities, University of Oslo (UiO), during the period 2021–2023. We would like to warmly thank OSEH for this support and all the members of this network for their invaluable contributions to the interdisciplinary conversations that have shaped the perspectives developed in this book: Stefan Helmreich, Jennifer Gabrys, Melody Jue, Tirza Meyer, Kimberley Peters, Bernhard Siegert, Franziska Torma, and Stefanie Hessler. The authors of this volume include many members of the network, alongside additional contributors.

We are grateful to the Media Aesthetics research group at the Department of Media and Communication (IMK) at the University of Oslo for their generous financial support of two book workshops held at the Lysebu Conference Center, one for network members in 2021 and another for the book's authors in 2022.

We also thank Henie Onstad Kunstsenter for supporting these workshops. In addition, the exhibition *The Atlantic Ocean. Art, Myths, Science*, presented at Henie Onstad in 2024, along with ongoing dialogues and collaborations with curators and artists working at the intersection of art, media, and oceanic environments, has contributed to perspectives developed in this book.

The development of this volume has also benefited from the opportunity to share work-in-progress and explore emerging ideas at several academic conferences. Among these were the NordMedia Conference in 2021, where the core group of the collaboratory organized the pre-constituted panel *Media Seas of the High North Atlantic*, and the NECS Conference in Oslo in 2023. These forums offered vital arenas for exchange and feedback that has informed the volume.

Our sincere thanks are extended to Tine Skjelstad for her invaluable research assistance on the book project in the fall of 2024 and to the Department of Media and Communication (IMK) for funding this support.

We further thank Oslo Metropolitan University (OsloMet) for providing meeting spaces and financial support for proofreading the book's introduction, and Kristiania University of Applied Sciences for generously making meeting rooms available throughout the entire process.

Contributors

Jørgen Alnæs, Department of Journalism and Media Studies, Oslo Metropolitan University, Norway

Kristin Asdal, TIK Centre for Technology, Innovation and Culture, University of Oslo, Norway

Camilla Brattland, The Arctic University Museum of Norway, Kutt, The Arctic University of Norway

Synne Tollerud Bull, School of Arts, Design, and Media, Kristiania University of Applied Sciences, Norway

Nancy Couling, Bergen School of Architecture, Norway

Liv Hausken, Department of Media and Communication, University of Oslo, Norway

Stefan Helmreich, Anthropology, Massachusetts Institute of Technology, USA

Melody Jue, Department of English, University of California, Santa Barbara, USA

Asko Lehmuskallio, Visual Studies Lab, Communication Sciences Unit, Tampere University, Finland

Susanna Lidström, Division of History of Science, Technology and Environment, KTH Royal Institute of Technology, Sweden

Tirza Meyer, Department of Culture, Religion, Asian and Middle Eastern Studies, University of Oslo, Norway

Fiona Polack, Department of English, Memorial University of Newfoundland and Labrador, Canada.

Laura Seddon, Marine Management Organisation, UK

Bernhard Siegert, Department of Media Studies Bauhaus-Universität Weimar, Germany

Nicholas R. Silcox, Department of English Colby College, USA

Philip Steinberg, Department of Geography Durham University, UK

Marie Stilling, TIK Centre for Technology, Innovation and Culture, University of Oslo, Norway

Susanne Østby Sæther, Henie Onstad Art Center, Norway

1 Introduction: Media Seas of the High North Atlantic

Jørgen Alnæs, Synne Tollerud Bull, Liv Hausken and Susanne Østby Sæther

The northern reaches of the North Atlantic Ocean, known in this book as the High North Atlantic, are undergoing profound and rapid transformations that threaten ecological balance and geopolitical stability. It is a maritime zone shaped by stark contrasts—where dark, cold winters meet the life-giving warmth of the Gulf Stream, and where summer's sunlit nights offer a fleeting yet vital burst of ecological productivity. As cold Arctic waters converge with warmer Atlantic currents, they sustain a unique—yet fragile—balance shaped by extreme seasonal variations. This delicate environment faces mounting pressures from human activity, including pollution, overfishing, and the accelerating impacts of climate change. Across the Atlantic, sea surface temperatures have reached record highs (Copernicus Climate Change Service, 2023; Met Office, 2023), driving shifts in species distributions as ecosystems respond to warming waters. Simultaneously, the Atlantic Meridional Overturning Circulation (AMOC), a critical regulator of the global climate system, shows alarming signs of weakening. These disruptions together signal a profound imbalance in Earth's planetary systems, with far-reaching consequences (Caesar et al., 2021; Ditlevsen and Ditlevsen, 2023). As the Arctic becomes increasingly Atlantified, with Atlantic waters encroaching further northward (Ingvaldsen et al., 2021), the ecological and geopolitical stakes across this part of the North Atlantic are reaching unprecedented levels.

These transformations intersect with intensifying geopolitical tensions. In 2024, NATO convened its first meeting of the Critical Undersea Infrastructure Network to address vulnerabilities in undersea cables, pipelines, and offshore wind farms—vital systems of critical infrastructure that underpin global communication, energy security, and the green transition (NATO, 2024). Recent incidents of potential sabotage in the Baltic Sea and near Lofoten, a rugged chain of islands extending into the Norwegian Sea above the Arctic Circle, underscore the fragility of these infrastructures in the face of escalating hybrid threats. The militarization of ocean governance, driven by rising tensions between state and non-state actors, adds further complexity to the already precarious balance between resource extraction, environmental sustainability, and global security. Greenland, officially recognized as the

DOI: 10.4324/9781003478805-1

world's largest non-continental island, situated at the center of the High North Atlantic, has again become the subject of territorial dispute. While Greenland maintains extensive self-governance, it remains an integral part of the Kingdom of Denmark. Proposals by the United States to acquire the island from Denmark have been formally rejected, leading the Danish monarch to underscore the Greenlandic symbol in his renewed royal coat of arms. By contrast, in his New Year's speech on January 1, 2025, the Prime Minister of Greenland called for full independence from Denmark. These dynamics reflect the escalating geopolitical tensions and international disputes that characterize the High North Atlantic, as nations vie for influence and control over this strategically significant region.

In parallel, industrial pressures on the seabed are intensifying. In 2024, Norway became the first nation in the world to open its ocean floors for deep-sea mining, transforming the abyssal plain from an inaccessible frontier into a battleground for resource exploitation. While framed under the rhetoric of the blue economy—promising sustainable growth and technological innovation—these initiatives often risk sliding into bluewashing, where industrial extraction is cloaked in claims of ecological responsibility. Proponents of deep-sea mining argue that marine minerals are essential to the green transition, particularly for renewable energy technologies, projecting substantial economic gains. Yet this logic of "green progress" relies on exploiting fragile deep-sea ecosystems, raising the paradox that one form of environmental protection may come at the cost of another. What appears as a win-win scenario—green and blue aligned—risks becoming a victory for extractivism, undermining the ecological values it claims to uphold. The seabed, long exploited for its oil and gas reserves in relatively accessible continental shelves, is now increasingly redefined as a site for deep-sea mining in more remote and vulnerable regions, raising distinct tensions around technological feasibility, corporate ambitions, and ecological stewardship, as well as issues of national sovereignty over these disparate areas.

Despite mounting pressures, the High North Atlantic remains a region of significant ecological value, hydrologically connected to distant seascapes through migratory patterns of both humans and various species, as well as shared feeding grounds that sustain a diverse and thriving ecosystem. This duality—of crisis and vitality—defines the High North Atlantic's pivotal role in global environmental and geopolitical systems. The convergence of cold, nutrient-rich Arctic waters with warmer Atlantic currents fosters biodiversity while also heightening the region's vulnerability to cascading ecological and climatic disruptions.

The northernmost parts of the North Atlantic also highlight a deeper cultural and existential challenge. Climate change represents more than a generically scientific or technical problem—it embodies the behavioral and ideological patterns that define much of the Global North. In this region, where ecosystems are particularly fragile, industrial and economic systems have historically prioritized short-term gains, often exacerbating environmental vulnerabilities. The accelerating loss of species and habitats underscores the need to reconsider humanity's relationship with the ocean—not as an inexhaustible resource, but as a dynamic and interconnected system that requires care, balance, and shared responsibility. Although the blue humanities have emerged to address oceanic crises through cultural and ethical inquiry, scholarship has largely overlooked this northern oceanic region, despite its critical ecological and geopolitical significance.

While earlier responses to environmental and climate challenges have been the domain of the natural and, to some extent, social sciences, the research field of blue humanities has developed since the early 2000s to illuminate the cultural, historical, and ethical dimensions of these crises. By examining the symbolic and material significance of the sea in literature, history, art, and philosophy, the field provides perspectives on how human relationships with the ocean have evolved and the ways in which they can be reimagined in the face of contemporary challenges.

Various scholars have developed and shaped the concept of blue humanities, which integrates the study of water, oceans, and maritime environments into the humanities. Initially growing out of Anglophone literary contexts during the first decade of the 2000s, one notable figure in the early formulation of this interdisciplinary field is Steve Mentz (2009). Exploring the cultural and literary significance of the sea, particularly in the early modern period, his work has brought attention to the intersections of humanity and the aquatic world. Others who have shaped the field from its early stage include Elizabeth DeLoughrey (2007, 2019), whose work examines the ocean's role in colonial histories, ecological crises, and decolonial futures. Hester Blum (2008, 2019) has similarly explored the ocean as a space of imaginative, material, and historical significance in maritime literature and, more recently, in writings on polar expeditions.

Often referred to as "the oceanic turn in literary studies" (Yaeger 2010, p. 524), the "cultural turn to the sea," and more formally introduced as "The Blue Humanities" (Gillis, 2013), as well as "blue cultural studies" (Mentz 2009) and occasionally "new thalassology" (from the Greek *thalassos*, the sea) (Mentz 2009), this turn integrates discourses from environmentalism, ecocriticism, maritime history, globalization, and postcolonialism, fostering a shift from terrestrial-centered frameworks to oceanic perspectives within the humanities.

Scholars across disciplines have challenged terrestrial-centered frameworks by emphasizing the ocean's fluidity, mobility, and spatial openness. In political geography, Philip E. Steinberg's *The Social Construction of the Ocean* (2001) and his later essay "Of Other Seas" (2013) argue for an ontological rethinking of the ocean—not merely as a backdrop or metaphorical space, but as a dynamic material entity with its own spatial and political logics. This perspective is further developed in his collaborations with Kimberly Peters, where they introduce the notion of a "wet ontology" (2015), later expanded into a “more than wet ontology” (2019) that includes ephemeral and unstable elements like mist and polar ice. Parallel developments in feminist environmental humanities similarly foreground the ocean’s material agency and ontological complexity. Stacy Alaimo integrates environmental philosophy, new materialism, science studies, and posthumanism to emphasize the entanglement of human and nonhuman worlds—particularly the ocean’s role in shaping ecological, cultural, and political relations. Astrida Neimanis (2012, 2017) further develops this line of thought with her concept of hydrofeminism, which highlights the ocean’s fluid materiality and embodied, relational ontologies across species and substances.

From an emphasis on Atlantic, Mediterranean, and Caribbean locations, the field and its objects of study have, since its first formulations around 2010, expanded to engage with global, non-Western, Pacific, and Indigenous materials and ideas. Mentz has recently described this earlier bias toward certain regions as an “Atlantic centrism” in *Blue Humanities* (Mentz, 2023). However, this alleged “Atlantic centrism” has primarily emphasized a strikingly limited part of the Atlantic oceans with less focus on the North Atlantic’s northernmost parts and the geographical north’s marginal regions. With this volume, we aim to further existing work in the blue humanities by expanding its geographical scope to include a more northern perspective, shedding light on the often-overlooked waters of the North Atlantic and the marginal regions of the geographical north. By doing so, we seek to supplement existing scholarship with insights into northern ecologies, histories, and cultural narratives, emphasizing their significance within global oceanic studies.

While the blue humanities have shed light on the cultural and ethical dimensions of oceanic crises, the field has yet to grapple fully with the role of technological mediation. Significant exceptions have been pivotal in the development of this book. Notably, as Stacey Alaimo points out: “While science studies have been vital for the environmental (green) humanities, it is even more essential for the development of the blue humanities, since most aquatic zones, species, and topics exist beyond human domains, requiring the mediation of science and technology” (Alaimo, 2019 p. 429). Despite this, much of the blue humanities discourse lacks a clear perspective on technological mediation in and of the ocean. With this volume, we aim to address this gap by examining how technological mediation shapes oceanic knowledge, perception, and interaction in the High North Atlantic.

Moreover, the ocean has increasingly been understood as a medium in itself—not merely a backdrop for technological mediation, but an active participant in communication, sensing, and world-making processes. Scholars such as John Durham Peters, Nicole Starosielski, and Melody Jue have contributed to this elemental media perspective, which foregrounds the role of natural elements in shaping media and mediation. In *The Marvelous Clouds* (2015), Durham Peters offers foundational insights into how the ocean functions both as an infrastructural conduit and an elemental force that conditions perception, organizes knowledge, and mediates relations between human and nonhuman life. Starosielski (2019) similarly emphasizes how environmental media infrastructures—such as undersea cables—mediate and transform ecological interactions, positioning the ocean itself as a medium among others. Building on this, Jue (2020) underscores that the deep ocean only becomes knowable through chains of mediation—remote sensing, imaging technologies, and technical systems that render otherwise inaccessible environments perceptible.

These perspectives accentuate the need for a critical media approach that understands mediation as infrastructural, elemental, and distributed across human and nonhuman relations. This volume draws on such insights from media theory, oceanic studies, and the blue humanities to examine how the High North Atlantic becomes visible, knowable, and computable through infrastructures, sensing systems, and multispecies entanglements. While these approaches are well established in some fields, they remain less familiar in others. The concept of Media Seas brings these strands together, offering a framework for understanding mediation not only as a process of transmission but as an elemental condition—shaping perception, experience, and epistemology.

Media Seas

Many have stated that our knowledge of the ocean is more shaped by technological mediation than other environments on Earth (see for example Lidström, Wickberg, and Gärdebo, 2023, p. 114). As noted above, scholars such as Alaimo and Jue have emphasized how the ocean's inaccessibility and fluidity demand mediation through science, technology, and representation.

Similarly, Lidström, Wickberg, and Gärdebo rightly note how governance, exploitation, and knowledge of the ocean are dependent on mediation processes. However, in this volume, we build on a more radical—in the sense of fundamental—concept of mediation. Mediation is not limited to hard-to-reach realms like the deep sea or outer space; all forms of sensing, knowledge, politics, and culture are mediated—not only technically, but also culturally and symbolically. Such a perspective has a longstanding tradition in media theory and has gained significant traction since the early 2000s. In the following, we will outline its key characteristics as they relate to this book and advance a critical media framework for the blue humanities—an approach we have termed Media Seas.

At the heart of this book lies this critical media perspective, which will be further elaborated throughout its chapters. This foundational perspective transcends the notion of a medium as a fixed entity, such as an object or apparatus, offering more profound insights into mediation as a dynamic process integral to perception, knowledge, and power. This shift represent a significant transition in perspective, which includes moving from considering media as mere collections of technologies to viewing mediums and media as concepts, ideas, and models that facilitate understanding of practices, articulations, and experiences.

Building on this foundational concept of mediation, this perspective emphasizes how matter and technology participate in shaping a critical and nuanced understanding of the environment.

This approach highlights the inherently relational nature of mediation and its indispensable role throughout human history. The idea that media shape our environments has a long history in media theory. Neil Postman (1970) notably framed media as ecological forces that structure perception and society. More recently, Lidström, Wickberg, and Gärdebo (2023) have developed the concept of environing media, drawing on theories of mediation and cultural techniques (*Kulturtechniken*) to examine how media practices constitute and condition environmental perception. Such a perspective inspires us to shift the focus away from media as a means of communication or material for aesthetic expression and toward the logic and processes of mediation. Again, it is vital to emphasize that mediation encompasses not only technical and material aspects but also forms, structures, models, patterns, symbols, languages, and more. As Raymond Williams (1977) articulated, a medium is not only the material of which it is composed but also includes social practices, a set of skills, habits, techniques, tools, codes, and conventions (see, e.g., Mitchell, 2008, p. 3). The media are our environment in the sense that our environment—material, technical, historical, cultural, societal, and elemental—mediates knowledge, experience, cognition, and perception.

A key perspective that underpins this volume is media aesthetics, an interdisciplinary field studying media technologies, sense perception, and mediation. It has evolved from earlier theoretical frameworks such as media ecology (Marshall McLuhan), media philosophy (Friedrich Kittler), medium theory (Joshua Meyrowitz), mediology (Régis Debray), and critical theory (notably Walter Benjamin). It is also influenced by ideas from new media (Mark Hansen) and visual culture (W.J.T. Mitchell). Media aesthetics emphasizes sensuous perception as culturally and historically situated and critically examines the pervasive technological mediation that defines contemporary society. This approach also explores how media infrastructures detect and process environmental signals, as seen in sensor-based mediation (Jennifer Gabrys), and how technical systems evolve and affect mediation through their material processes (Gilbert Simondon).

In addition to offering a fundamental perspective on mediation, this volume draws upon media theoretical approaches that emphasize the constitutive role of material and operational practices. Central among these is the concept of *cultural techniques* (*Kulturtechniken*), which shifts attention from symbolic content to the technical, ritual, and infrastructural operations that underpin cultural formations. This approach is particularly associated with the work of Bernhard Siegert, a contributor to this volume, whose research examines how media practices organize perception and enact social and epistemic orders. This emphasis on operational and material processes resonates with one influential strand of media archaeology, particularly its focus on the technical conditions of inscription and storage that underlie media systems. Friedrich Kittler (1999) and Wolfgang Ernst (2013) are central figures in this approach, which Jussi Parikka (2015) extends toward ecological and geological materialities and temporalities. This lineage of media archaeology emphasizes the materiality and temporality of media infrastructures, drawing on Kittler's notion of "inscription systems" (*Aufschreibesysteme*) to foreground the technical and procedural conditions through which information is recorded, stored, and transmitted.

This strand of media archaeology aligns with the aims of this volume by investigating the material foundations of media technologies and their archival dimensions. Expanding this focus beyond human-made archives, the project also considers natural archives—oceanic records embedded in ice cores, deep-sea sediments, and marine organisms. For example, the inner eye lens of the notoriously long-lived Greenland shark can be carbon-dated, revealing environmental events over centuries. These natural archives document planetary change across deep time, bridging media-archaeological approaches with environmental humanities and emphasizing how both technological and natural systems mediate cultural memory.

Finally, the critical media perspective in this volume ties into Elemental Media, a research area gaining prominence within the blue humanities, media studies, and ecological criticism. Notable figures such as John Durham Peters, Nicole Starosielski, Melody Jue, and Stacy Alaimo, along with Jussi Parikka and Lisa Parks are key contributors to these explorations. Elemental Media gathers interdisciplinary research on elements like earth, water, air, chemicals, minerals, fuels, and plastics, tracing these materials—which constitute and permeate human life and our environment—through the practices that both shape and are shaped by them. By examining the origins, provenance, movement, transformation, and impact of these elemental media, scholars aim to understand their role in human and nonhuman interactions.

This volume builds on a critical media perspective with relevance far beyond the ocean. Yet here, we apply this perspective specifically to the ocean as a site that both challenges and is challenged by ecological, technical, political, and cultural conditions. The term Media Seas introduces a theoretical framework for exploring the entanglements between media technologies, cultural representations, and maritime environments. It encourages interdisciplinary dialogue and offers deeper insight into the ecological, cultural, and technological dimensions of the ocean. By tracing how media mediate and are mediated by oceanic conditions, Media Seas helps illuminate the political and ecological challenges of some of the planet's most complex and least understood environments.

The phrase Media Sea deliberately gestures toward a fluid semantic field: not only seas as media—material, elemental, and infrastructural—but also media as that which *sees*, perceives, and renders the ocean sensible. This play on Media Seas echoes Nicole Starosielski's (2015) exploration of undersea cable infrastructures as both conduits of signal and agents of geopolitical vision—a perspective further developed by Nicholas R. Silcox in this volume, and resonant with Melody Jue's (2020) theorization of "wild blue media," which, among other interventions, recalibrates perception through immersion. Media Seas also drifts toward *mediacy*, blurring the boundary between what media *is* and what it *does*—between medium as substance and mediation as process.

Seas of the Greenland Shark

On August 16, 1858, midway between Ireland and Newfoundland, two ships positioned themselves stern to stern and connected two enormous rolls of telegraph wire. Then, the two ships continued in opposite directions, to the west and to the east, while lowering the cables onto the uncharted depths of the ocean. Approximately one week later, the ships reached land on each side of the North Atlantic, making it possible for words to travel from Europe to America in a matter of minutes rather than weeks. At this moment, by the first cable across the Atlantic Ocean, the world was made smaller. The Austrian author Stefan Zweig calls it one of the "*Sternstunden der Menschheit*" as a decisive moment in the history of humanity: "By their own creative power, humanity now lives simultaneously from one end of the Earth to the other, like an omnipresent deity" (Zweig [1940]1979, p. 129, our translation). We can understand this sentiment in light of Marshall McLuhan's assertion that media are extensions of the human sensory apparatus and that the "message" of any medium or technology is "the change of scale or pace or pattern that it introduces into human affairs" (McLuhan [1964] 2003, p. 20). The first undersea telegraph cable radically changed the communicative distance between people who were still physically far apart.

As Nicole Starosielski notes, such important—but, in many senses, invisible—submarine cables have not received the cultural historical attention they deserve "due to a broader social tendency to overlook the distribution of modern communications in favor of the more visible processes of production and consumption" (2015, p. 4). Communication cables are just one of many connections running through the northernmost North Atlantic, constantly carrying data, images, and ideas. In addition, currents, people, and animals also traverse this part of the ocean. The following discussion will examine the movement of these latter three factors more closely.

Beginning with the movement of seawater, surface currents, driven by wind and salinity, carry warm water northward, while deep currents transport cold water southward. This current system, known as the AMOC, significantly influences the climate in the Atlantic Ocean, transporting energy and warmth and rendering the coastal climate around the High North Atlantic warmer than it would otherwise be. Although recent reports indicate a slowdown in these currents, and a total collapse is considered unlikely (Ditlevsen and Ditlevsen, 2023), the potential consequences of such changes highlight the critical importance of the AMOC in maintaining climate stability. Such scenarios form the backdrop of the 2004 American science fiction disaster film

The Day After Tomorrow, depicting catastrophic climatic effects following the disruption of the North Atlantic Ocean circulation. In recent decades, warm, salty water flowing northwards from the Atlantic Ocean has affected the climate in the Arctic Ocean, a process known as atlantification. This has led to melting sea ice entailing that species from the Atlantic Ocean establishing themselves further north, resulting in significant changes in the ecosystems of the Arctic (Ingvaldsen et al., 2021).

The movement of seawater has been depicted as a conveyor belt, a metaphor first introduced by geochemist Wallace Broecker to describe the ocean's role in transporting warmth across the planet (Broecker, 1991). Complementing this image, the ocean is also perceived as an undulating network of pathways and routes for travel. For centuries, people have taken the sea route across the High North Atlantic. Around the year 1100 CE, ancestors of the modern Inuit traveled from Canada by sea to inhabit Greenland. Numerous accounts suggest their journeys extended to the Orkney Islands north of Britain. While these narratives remain unverified (Trynkina, 2014; Westaway, 2020), they attest to the enduring belief in the sea as a unifying pathway. Similarly, accounts of St. Brendan's legendary voyages from Ireland to the far northern and western reaches of the Atlantic Ocean—possibly reaching Newfoundland in the 6th century CE—underscore the sea as a pathway connecting the countries surrounding the High North Atlantic. Archaeological evidence has now confirmed what the Norse sagas long described: Viking explorers led by Leif Erikson reached Newfoundland around 1000 CE. The discovery of the Norse settlement at L'Anse aux Meadows resolved the longstanding question of whether Vikings had indeed crossed the entire North Atlantic as claimed in their historical accounts (Wallace 2003).

These northern maritime areas have been vital waterways for both travel and transporting goods, not only for people in the countries bordering the High North Atlantic; they also functioned as key transit paths for nations beyond these regions. This role will become increasingly significant in the years to come as the melting ice is expected to open up transpolar sea routes connecting the North Atlantic and the Pacific Ocean, positioning the High North Atlantic as a crucial gateway (Brady, 2022). For decades, vessels have navigated along its north-south axis. While Russia signals ambitions to scale up its presence near the Norwegian border, China's *Polar Silk Road* initiative envisions the Arctic as a vital corridor for future trade and resource extraction, despite ongoing environmental concerns and logistical setbacks (Lamazhapov, Stensdal, and Heggelund, 2023). During the Cold War, the GIUK gap—located between Greenland, Iceland, and the UK—held strategic importance as it was the only route available for Soviet submarines to move from the Barents Sea to the Atlantic Ocean and onwards to the eastern coast of the USA. This significance was instrumental in the establishment of the NATO station at Keflavik, Iceland, where submarine surveillance was a primary mission (NATO, n.d.). The importance of the GIUK gap is vividly depicted in the 1990 American submarine film *The Hunt for Red October*, arguably the most recognized portrayal of the deep waters of the High North Atlantic.

Beside military and strategic transport, people have traversed the High North Atlantic in pursuit of the ocean's rich resources. In *The Mortal Sea*, W. Jeffrey Bolster documents how Europe's coastal waters were depleted over centuries, resulting in dramatic declines in fish abundance, size, number, and diversity. This depletion drove a large number of fishermen across the High North Atlantic to the rich fishing grounds off Newfoundland, where they encountered marine ecosystems reminiscent of their own waters from generations past. These voyages represented more than just geographic traversals. As Bolster argues, they constituted journeys through "ecological time" to reclaim the riches of the once-bountiful sea that their ancestors had known (Bolster, 2012, p. 45).

Having explored the movement of water and people across the High North Atlantic, attention now turns to the third category: the dynamic movement of the diverse and vibrant marine life inhabiting these waters. The intricate ecosystems of the northernmost parts of the North Atlantic are home to a diverse multitude of creatures, from the smallest plankton to majestic whales, each playing a crucial role in the health and balance of these marine environments. The movements and interactions of these living beings not only reflect the dynamic nature of the oceans but also underscore the interconnectedness of oceanic ecosystems and the broader environmental challenges they face.

Some species are uniquely adapted to the High North Atlantic as their habitat. One such species is the Atlantic salmon, which hatches in rivers along both sides of the High North Atlantic, from Maine and northwards along the coast of Canada, from Spain and up to Russia, and from the British Islands up to Iceland and Greenland. From their river hatcheries, the salmon migrate into the sea; satellite tags have revealed that salmon can travel vast distances and even that salmon from both sides of the ocean can sometimes roam in the same areas (Rikardsen et al., 2021).

The High North Atlantic is also home to one of the world's longest living creatures. The Greenland or gurry shark's habitat spans from the British Isles to Newfoundland, extending north to the Arctic, including Baffin Bay, the Norwegian Sea, and parts of the Barents Sea (see Figure 1.1). The shark is featured in Inuit and Canadian folklore, notably appearing in the Sedna myth (see e.g., Nielsen, 2017). In Iceland, fermented and dried Greenland shark (*kæstur hákarl*) is known as a delicacy. Within the last decade, the shark has gained cultural prominence in Norway, featuring in author Morten Strøksnes' award-winning literary non-fiction book,[1] which has also been adapted to a successful theater play. These diverse cultural references illustrate how the Greenland shark has inspired a range of ideas and values across multiple cultural contexts, geographic regions, and historical periods. We highlight this species here because its range coincides with our study's geographic focus. However, our attention to the Greenland shark serves a broader purpose—it gestures toward the interconnected species that inhabit these waters, emphasizing that our study concerns not just the geographical regions but the ecological and cultural relationships it fosters.

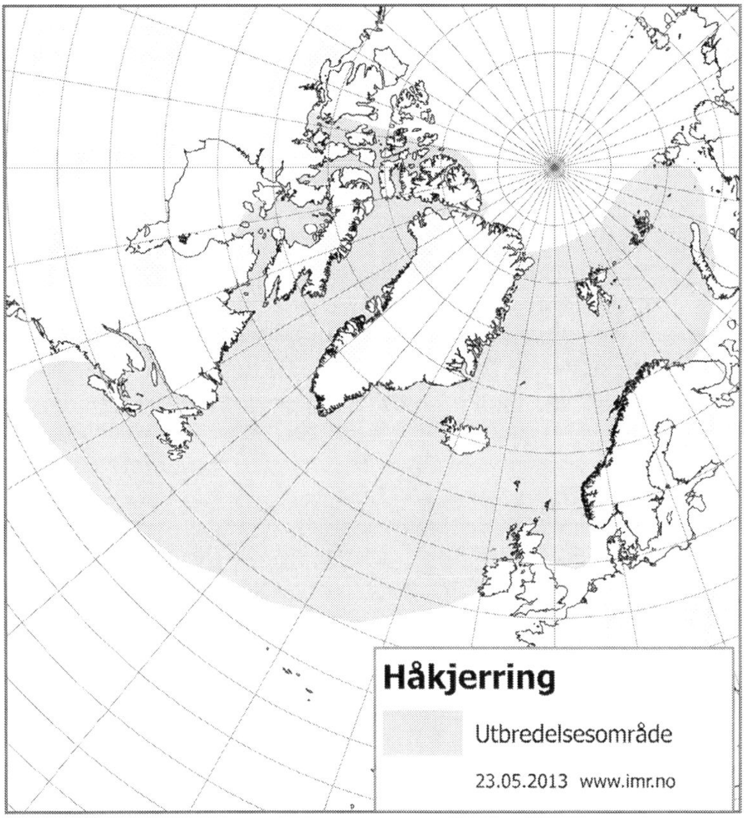

Figure 1.1 Area of distribution of the Greenland shark ("Håkjerring")
Source: Institute of Marine Research, Norway

"Seas of the Greenland Shark" could have been incorporated as a subtitle for this book project, effectively highlighting the specific oceans that are the focus of this study through the shark's distribution area. Indeed, marine biologist Julius Nielsen has suggested that a more appropriate name for the species could be the "Northern North Atlantic Deep-Sea Shark" to more precisely indicate its range (Nielsen in conversation with Kølbæk Iversen in exhibition catalogue, *Rovhistorier*, 2022 [2025]). However, for this volume, we propose "The High North Atlantic" as a more evocative term for the oceans under exploration. High North Atlantic is a compound term that brings together the established designations of the expansive "North Atlantic" and the "High North." The latter is a well-recognized English term corresponding to the Norwegian "*Nordområdene*" (literally "the northern areas"), commonly used in juridical and governmental contexts.

An alternative designation to "High North Atlantic" could be "Subpolar North Atlantic," but this and similar constructions tend to emphasize the polar aspect of the North Atlantic maritime region. The focus on the northern polar region draws attention away from the much less studied oceans that this book project aims to explore. Despite the prosperity and economic wealth that many associate with the Global North, this area is more locally—and in the relevant countries—associated with challenges such as barren land and harsh climate, periphery related to the center, challenges of depopulation as people increasingly move to cities, challenges regarding border politics and divisions of cultures across nations, as well as colonial and Indigenous perspectives. These complexities are exemplified by the current situation in Greenland, where questions of political autonomy, resource sovereignty, and Indigenous self-determination remain central to both domestic policy and international negotiations (Nuttall, 2022).

In recent decades, the notion of *Thule* has resurfaced as an allusion to the mythic conceptions of the High North Atlantic. This idea draws upon "Ultima Thule," a term historically used to signify a distant, often mythical place at the edge of the known world (Jones and Symmes, 2010). Its origins trace back to ancient Greek and Roman geography and mythology. The Greek explorer Pytheas of Massalia (4th century BCE) was the first to use the term, describing a land he claimed to have visited far north of Britain. He depicted it as a place where the sun barely sets in the summer, possibly referring to the Arctic, Iceland, Norway, or the Shetland Islands (Burton, 1875; Hastrup, 2007; Stefansson, 2022).

The Norwegian explorer and diplomat Fridtjof Nansen sought to anchor the myth of Thule to national geography, declaring in 1911 that "every account of Thule fits Norway" (Nansen, 1911, p. 46). His claim positioned Norway not only as a northern frontier but as a land imbued with the mystery, extremity, and imaginative projection historically associated with Ultima Thule. Yet the concept proved mobile. When the US Air Force established a base in northwestern Greenland in 1951, it named the installation the Thule Air Base (renamed Pituffik Space Base in 2023), symbolically reinscribing this liminal Arctic region with the same mythic associations. The term continues to haunt contemporary imaginaries of the North; Danish photographer Henrik Saxgreen, for instance, uses it in the title of his 2014 book on Greenland, evoking a place still perceived as remote, enigmatic, and beyond the limits of ordinary experience.

In the Roman world, Thule (or *Ultima Thule*, meaning "farthest Thule") came to symbolize the farthest point of exploration, a boundary beyond which the world was unknown or unknowable. It was used metaphorically to represent the unattainable or the limits of human knowledge and imagination. Today, the notion of Thule is used as an allusion to the old mythical conception of the High North Atlantic in several brands, such as a cartoon, a musical group, a choir, a record company, a travel agency, an equipment manufacturer, and so on. Although people's movements in this region have long been driven by both survival and the pursuit of wealth, the High North Atlantic today is notably prosperous, and much of its connectivity and mediation are enabled by this wealth. Just as the first telegraph cable described above was sponsored by millionaire interests, the mapping of deep-sea valves, waves, ice, and salmon has been carried out by economically powerful states. The region is rich in natural resources yet accessing them demands vast financial investment due to the extremely difficult physical conditions. In this sense, one kind of wealth—natural abundance—relies on another: the economic capacity to extract and mobilize it.

The high cost of oceanic exploration has also fostered strategic collaborations between industry and research, where joint expeditions and data-sharing agreements help map seabed resources while generating other forms of scientific knowledge (see Jue, this volume). The relationship between climate and business development is particularly precarious in the northern part of the North Atlantic. The oil industry here is mainly offshore, with oil and gas installations on the seabed, pipelines, floating facilities, and ship traffic. This industry is undergoing a so-called green shift, and new industries are emerging, such as power production from offshore wind and the development of seabed mining. As Polack and Farquharson observe, "The continuing pursuit of oil and gas … constitutes an extreme yet highly revealing example of the lengths to which corporations and governments have been prepared to go in order to maintain the centrality of fossil fuels" (2022, p. 3).

Through its focus on "Media Seas of the High North Atlantic," this book aims to combine dimensions of *ocean, north*, and *media*. What sets these oceans apart is their depth, darkness, temperature variations and ecological significance. The concept of Media Seas becomes particularly relevant here. It emphasizes the role of technical, cultural, economic, and political mediation of oceans and bodies of water that are both rich and vulnerable, ecologically and geopolitically, and are currently undergoing significant changes. This critical media approach enables complex and more comprehensive explorations of how the oceans are exploited, which voices are expressed, and which narratives are emphasized. As stated earlier, the term Media Seas points to a two-fold process: not only seas as material, elemental, and infrastructural media, but also media as that which sees, perceives, and renders the ocean sensible. This understanding of "media sees" resonates with Nicole Starosielski's (2015) account of submarine cable infrastructures as both carriers of signal and agents of geopolitical vision, and also aligns with Melody Jue's (2020) theorization of "wet media," which recalibrates perception through immersion. Media Seas also leans toward mediacy, unsettling the boundary between what media is and what it does—between medium as substance and mediation as process.

The importance and challenges of cross-disciplinary collaboration in addressing pressing societal and ecological issues cannot be overemphasized. *Media Seas and the Blue Humanities: Studies of the High North Atlantic* exemplifies such an approach, bridging various fields within the humanities to extend their relevance to contemporary challenges. By integrating insights from media and communication studies with human geography, anthropology, environmental studies, cultural and technological history, as well as visual studies, architecture, art, and literature, this volume offers a multifaceted, critical media perspective on the High North Atlantic.

The breadth of scholarly contributions enriches the Media Seas concept, bringing diverse perspectives to the dynamics of this unique maritime environment. As a pivotal intervention, the anthology advances environmental and blue humanities in dialogue with media studies, offering fresh theoretical and methodological tools for exploring the northernmost parts of the North Atlantic's increasingly contested ocean space in the 21st century. Ultimately, it underscores the humanities' vital role in navigating and addressing global challenges.

Chapters

Each chapter in this volume contributes uniquely to the overarching project by examining different facets of the Media Seas concept. By focusing the media lens on specific oceanic realms, the project seeks to uncover the unseen, unheard, and untouched elements in and of these northern oceans, providing fresh perspectives on the complex mediations within and surrounding the High North Atlantic.

The book is structured into three parts, loosely following the vertical movement of the water column—from the substance of the ocean itself to its surface and deep-sea ecologies. Each part introduces a different spatial orientation while remaining attentive to the transscalar and interconnected nature of oceanic media. *Of the Sea* explores the materiality and mediations of seawater, plankton, and the blue economy. *On the Sea* shifts focus to the surface and upper layers, where navigation, architecture, sea ice, and remote sensing coalesce. Finally, *In the Sea* investigates the seabed, coral ecosystems, Indigenous infrastructures, and speculative futures, revealing how depth operates not only as geography but as affect, history, and political terrain.

From plankton filters and rogue waves to hydrothermal vents and Greenland sharks, the chapters that follow invite readers into a dynamic encounter with the ocean's forms, figures, and infrastructures. Together, they map out a media-ecological landscape of the High North Atlantic—an oceanic region shaped as much by imagination, technology, and representation as by climate, currents, and capital. Rather than offering conclusive answers, this volume opens a space for ongoing inquiry, urging readers to engage critically and creatively with the mediated oceans of our time. The chapters that follow explore this region through distinct yet interrelated case studies. Moving from surface to seafloor, from scientific instruments to Indigenous cosmologies, from imagined futures to entangled histories, they demonstrate how the sea's depth and complexity emerge through media practices.

Of the Sea begins with investigations into how the sea is made knowable through technical instruments and representational forms. Bernhard Siegert's *The MediaNature of Seawater* reframes plankton research as a history of media, tracing how filters, sensors, and visualization technologies constitute different "medianatures" of marine life. Marie Stilling and Kristin Asdal follow, by shifting the focus to documents—grant applications, reports, and white papers—as key mediators of the emerging seaweed economy. In *Moving the Green Economy to the Blue Sea*, they demonstrate how documents function not just as representations, but as sites of transformation where economic and ecological visions take hold. Stefan Helmreich's chapter on *The Draupner Wave* continues this media-historical mode, exploring how a rogue wave becomes a mythic, scientific, and cinematic event that transforms understandings of oceanic unpredictability. Finally, Nancy Couling's *Architectural Mediation of the Norwegian Sea* considers how architecture and spatial design practices can reorient perceptions of oceanic territory, calling for new forms of engagement with the sea as both a lived and imagined environment.

In *On the Sea*, the emphasis shifts to surface conditions and mobility. Fiona Polack's *Oil Media on the Grand Banks* explores how oil platforms act as infrastructural and epistemic nodes, shaping the Northern maritime environment through architectural and urban imaginaries. Asko Lehmuskallio's *Charting Courses* turns to navigation and the visual regimes enabled by AIS (automated identification systems), showing how satellite-based infrastructures alter the way seafarers and states perceive and govern the sea. Laura Seddon and Phil Steinberg's *Sea Ice Extent and the Sea Ice Index* examines the measuring and mediating of Arctic sea ice as a climate indicator, showing how satellite data both simplifies and obscures the complex materialities and politics of ice.

In the Sea brings us into the depths—biological, political, and conceptual. Tirza Meyer and Susanna Lidström's *Marvelous Deposits* considers the naming of hydrothermal vents as a form of oceanic mediation that both reflects and advances extractivist imaginaries. Jørgen Alnæs' *Territorializing the Seabed through the Battle of Deep-Sea Mining* examines how Norway's continental shelf becomes "media territory" through journalistic storytelling, musical performance, and naming practices, situating seabed mining within broader national imaginaries. Nicholas R. Silcox's *Linking to Sedna* explores how the SednaLink subsea cable project envisions infrastructure differently, rooted in Inuit cultural contexts and offering a counter-narrative to colonial media regimes.

The final chapters examine speculative futures and nonhuman perspectives. Camilla Brattland and Synne Tollerud Bull's *Ocean Futures in the Making* analyzes Ursula Biemann's *Acoustic Ocean*, where Sámi cosmology and sensing technologies converge in a multispecies meditation on extractivism, resilience, and Indigenous futurism—invoking the erased coastal Sámi presence in Northern Norway. Susanne Østby Sæther's chapter *In the Eyes of the Greenland Shark* discusses Marie Kølbæk Iversen's immersive video installation based on microscopic recordings of the shark's inner eye lens, reframing its eye lens as a temporal archive distributed between biological, machinic and elemental agents. Closing the volume, Melody Jue's *Cold-Water Coralations* contrasts warm-water coral imaginaries with the sonic and geopolitical entanglements of cold-water corals, particularly *Lophelia pertusa*, which resist photographic representation and dwell near petroleum infrastructures, complicating assumptions about ecological visibility and proximity.

Together, these chapters offer a multifaceted engagement with the High North Atlantic, foregrounding the vital role of media in shaping ecological knowledge, geopolitical claims, and cultural imaginaries in this region. Rather than prescribing definitive interpretations, this book invites readers to think with the sea—to explore, question, and extend the connections it traces across disciplinary boundaries and oceanic depths. In doing so, it opens space for further research into the mediated ocean worlds that shape, and are shaped by, our collective futures.

Note

1 Translated to English as *The art of catching a large shark from a tiny rubber dinghy in a big ocean* (2018).

References

Alaimo, S. (2019). Introduction: Science studies and the blue humanities. *Configurations*, 27(4), 429–432. doi:10.1353/con.2019.0028.

Blum, H. (2008). *The view from the masthead: Maritime imagination and Antebellum American sea narratives.* University of North Carolina Press.

Blum, H. (2019). *The news at the ends of the earth: The print culture of polar exploration.* Duke University Press.

Brady, A.-M. (2022, October 24). China's polar silk road: A long game or a failed strategy? *The Arctic Institute.* https://www.thearcticinstitute.org/china-polar-silk-road-long-game-failed-strategy/.

Bolster, W.J. (2012). *The mortal sea: Fishing the Atlantic in the age of sail.* Belknap Press of Harvard University Press.

Broecker, W. (1991). The great ocean conveyor. *Oceanography*, 4(2), 79–89, doi:10.5670/oceanog.1991.07.

Burton, R.F. (1875). *Ultima Thule: Or, a summer in Iceland* (Vol. 1). William P. Nimmo.

Caesar, L., Rahmstorf, S., Robinson, A., Feulner, G., and Saba, V. (2021). Observed fingerprint of a weakening Atlantic Ocean overturning circulation. *Nature Climate Change*, 11(7), 620–626. doi:10.1038/s41586-018-0006-5.

Copernicus Climate Change Service. (2023, August 8). Global sea surface temperature reaches a record high. https://climate.copernicus.eu/global-sea-surface-temperature-reaches-record-high.

DeLoughrey, E.M. (2007). *Routes and roots: Navigating Caribbean and Pacific Island literatures.* University of Hawaii Press.

DeLoughrey, E.M. (2019). *Allegories of the Anthropocene.* Duke University Press.

Ditlevsen, P.D. and Ditlevsen, S. (2023). Warning of a forthcoming collapse of the Atlantic meridional overturning circulation. *Nature Communications*, 14(1), 4254. doi:10.1038/s41467-023-39810-w.

Gillis, J.R. (2013, May/June). The blue humanities. *Humanities*, 34(3). https://www.neh.gov/humanities/2013/mayjune/feature/the-blue-humanities.

Hastrup, K. (2007). Ultima Thule: Anthropology and the call of the unknown. *Journal of the Royal Anthropological Institute*, 13(4), 789–804. doi:10.1111/j.1467-9655.2007.00460.x.

Ingvaldsen, R.B., Assmann, K.M., Primicerio, R., Fossheim, M., Polyakov, I.V., and Dolgov, A.V. (2021). Physical manifestations and ecological implications of Arctic Atlantification. *Nature Reviews Earth & Environment*, 2(12), 874–889. doi:10.1038/s43017-021-00228-x.

Jones, D., and Symmes, J.C. (2010). Ultima thule: Arthur Gordon Pym, the polar imaginary, and the hollow earth. *The Edgar Allan Poe Review*, 11(1), 51–69.

Jue, M. (2020). *Wild blue media: Thinking through seawater*. Duke University Press.

Lamazhapov, E., Stensdal, I., and Heggelund, G. (2023). China's polar silk road: Long game or failed strategy? (November 14, 2023). *The Arctic Institute*. https://www.thearcticinstitute.org/china-polar-silk-road-long-game-failed-strategy/.

Lidström, S., Wickberg, A., and Gärdebo, J. (2023). Ocean environing media: Datafication of the deep sea. In A. Wickberg, andJ. Gärdebo (Eds.), *Environing media* (pp. 114–133). Routledge.

McLuhan, M. (1964). *Understanding media: The extensions of man*. McGraw-Hill.

Mentz, S. (2009). Toward a blue cultural studies: The sea, maritime culture, and early modern English literature. *Literature Compass*, 6(5), 997–1013.

Mentz, S. (2023). *An introduction to the blue humanities*. Routledge.

Met Office. (2023, June 16). Sea surface temperatures breaking records. https://www.metoffice.gov.uk/blog/2023/sea-surface-temperatures-breaking-records.

Mitchell, W.J.T. (2008). Addressing media. *MediaTropes*, 1(1), 1–18.

Nansen, F. (1911). *Nord i Tåkeheimen: Utforskingen av jordens nordligste strøk i tidligere tider*. Dybvad.

NATO. (2024, May 23). NATO holds first meeting of critical undersea infrastructure network. *NATO*. https://www.nato.int/cps/en/natohq/news_225582.htm.

NATO. (n.d.) *Iceland and NATO*. https://www.nato.int/cps/en/natohq/declassified_162083.htm.

Neimanis, A. (2012). Hydrofeminism: Or, on becoming a body of water. In H. Gunkel, C. Nigianni, and F. Söderbäck (Eds.). *Undutiful daughters: Mobilizing future concepts, bodies and subjectivities in feminist thought and practice* (pp. 85–99). Palgrave Macmillan.

Neimanis, A. (2017). *Bodies of water: Posthuman feminist phenomenology*. Bloomsbury Academic.

Nielsen, J. (2017). *The Greenland shark (Somniosus microcephalus) diet, tracking and radiocarbon age estimates reveal the world's oldest vertebrate* [PhD thesis]. University of Copenhagen.

Nielsen, J., and Kølbæk Iversen, M. (2025). Merman. Greenland shark. North Atlantic deep-sea shark: Julius Nielsen in conversation with Marie Kølbæk Iversen. In Overgaden (Ed.), Rovhistorier [Exhibition catalogue]. Overgaden. https://overgaden.org/en/publications/rovhistorier.

Nuttall, M. (2022). Wild lands, remote edges: Formations and abstractions in Greenland's resource zones. In A. Mason (Ed.), *Arctic abstractive industry* (pp. 83–107). Berghahn.

Peters, J.D. (2015). *The marvelous clouds: Toward a philosophy of elemental media*. University of Chicago Press.

Polack, F., and Farquharson, D. (2022). *Cold water oil: Offshore petroleum cultures*. Routledge.

Postman, N. (1970). The reformed English curriculum. In A.C. Eurich (Ed.), *High school 1980: The shape of the future in American secondary education* (pp. 160–168). Pitman.

Rikardsen, A.H., Righton, D., Strøm, J.F., Thorstad, E.B., Gargan, P., Sheehan, T. et al. (2021). Redefining the oceanic distribution of Atlantic salmon. *Scientific Reports*, 11, Article 12266. doi:10.1038/s41598-021-91137-y.

Saxgren, H. (2014). *Ultima Thule: Grønland ved verdens ende.* Gyldendal.

Siegert, B. (2015). *Cultural techniques: Grids, filters, doors, and other articulations of the real.* Fordham University Press.

Siegert, B. (2024). *Final frontiers: Eine Medienarchäologie des Meeres.* Brill/Fink.

Steinberg, P.E. (2001). *The social construction of the ocean.* Cambridge University Press.

Steinberg, P.E. (2013). Of other seas: Metaphors and materialities in maritime regions. *Atlantic Studies*, 10(2), 156–169.

Steinberg, P.E., and Peters, K. (2015). Wet ontologies, fluid spaces: Giving depth to volume through oceanic thinking. *Environment and Planning D: Society and Space*, 33(2), 247–264.

Steinberg, P.E., and Peters, K. (2019). Cross-currents and undertows: A response. *Dialogues in Human Geography*, 9(3), 333–338. https://doi.org/10.1177/2043820619879521.

Starosielski, N. (2015). *The undersea network.* Duke University Press.

Starosielski, N. (2019) The elements of media studies. *Media+Environment*, 1(1). https://doi.org/10.1525/001c.10780.

Stefansson, V. (2022). *Ultima Thule: Further mysteries of the Arctic.* DigiCat.

Strøksnes, M. (2018). The art of catching a large shark from a tiny rubber dinghy in a big ocean. Vintage.

Trynkina, D.A. (2014). Kayakers near Scotland's northern shores at the turn of the 17th–18th centuries: Main theories of origin. *Archaeology, Ethnology and Anthropology of Eurasia*, 42(1), 115–122. doi:10.1016/j.aeae.2014.10.011.

Wallace, B. (2003). The Norse in Newfoundland: L'Anse aux Meadows and Vinland. Newfoundland Studies, 19(1), 5–43.

Westaway, J. (2020). The Orkney finmen legends: From early modern science to modern myth. *The Public Domain Review.* http://clok.uclan.ac.uk/35503/.

Williams, R. (1977). *Marxism and literature.* Oxford University Press.

Yaeger, P. (2010). Sea trash, dark pools, and the tragedy of the commons. *PMLA*, 125 (3), 523–545.

Zweig, S. (1979). *Sternstunden der Menschheit.* Fischer Taschenbuch Verlag.

2 The MediaNature of Seawater

Filtering, Counting, Scanning, Visualizing, and Digitizing the Plankton of the North Atlantic

Bernhard Siegert

Introduction

It is a commonplace that knowledge about the ocean, especially the deep sea, depends on technological mediation more than any other environment on Earth. "The importance of mediation here is key," writes Melody Jue (2020) in *Wild Blue Media*, "in order to study the ocean—especially the deep ocean—scientists need a variety of instrumentation, satellites, remotely operated vehicles (ROVs), submersibles, sonar, and other technical prostheses for sampling and sensing" (p. 3). The nature of the ocean thus studied, however, is far from being a stable ontological category; rather, it is inextricably entangled with media practices and cultural techniques that generate the data which stand in for the natural phenomena being studied.

This chapter undertakes to recontextualize the history of plankton research in the North Atlantic within the framework of a praxeology of media, or, more precisely, of a theory of cultural techniques. "Media have," according to John Durham Peters (2015), "shifted from mass media to cultural techniques not only in theory, but in fact" (p. 325). The media concept applied in the study of cultural techniques goes beyond a representational concept of media. Instead of thinking of "media" as the technical means of communicating meaning, media are here understood as "environing" cultural techniques, and not just as "semiotic inputs into people's heads" (Peters, 2015, p. 4). Media in this sense are "world enabling infrastructures" or "infrastructures of being" (Peters, 2015, p. 15). If the object of all microbiological oceanography in any form is always microbial seawater, then a follow-up to Peters' take on media as infrastructures of being would be that there is no seawater as such; rather, only cultural techniques of sampling, recording, scanning, imaging, and databasing. Seawater is a "nature-culture," as Stefan Helmreich (2011) has argued, deploying a term that was first used by Marilyn Strathern and later by Donna Haraway (p. 132). However, since I argue less in terms of anthropology and more in terms of media history and the theory of cultural techniques, I prefer Jussi Parikka's term "medianature" (Angerer, 2020; Parikka, 2012). But the medianature of seawater is not exclusively determined by the applied method of observation and extraction. It is as well determined by an

DOI: 10.4324/9781003478805-2

assemblage of scientific discourse, the geopolitical and economic interests of state agencies, and the interests of extraction capitalism. The discourse which is empowered by this assemblage sets the rules that usually determine what to measure, what to collect, and where and how to collect it. What interests me is how morphologists and evolutionists, marine physiologists, and microbiologists encounter the ocean as a space to be determined by distinctions—distinctions based on decisions taken a priori by the respective disciplines, their methods, and the instruments and practices they apply. "[Sea]water is material-discursive," as curator Stefanie Hessler has put it with reference to Karen Barad (Hessler, 2019, p. 215).

To accomplish this turn from a representational to an infrastructural concept of media, I propose a media-theoretical and media-historical twist to the notion of "environing technologies" introduced by environmental historian Sverker Sörlin and historian of science Nina Wormbs. By the term "environing technologies" Sörlin and Wormbs understand "the practices whereby humans make environments, both by shaping material environments and by bringing into being perceived or conceptual understandings of the environment or different aspects of it" (Sörlin and Wormbs, 2018, p. 106). Lidström, Wickberg, and Gärdebo (2023) have taken a first step in interpreting the concept of environing technologies in media-theoretical terms: "We understand environing to mean the process of both knowing and changing the environment, and environing media to mean the technical means by which this process can take place" (p. 115).[1] This shifts the agency that Sörlin and Wormbs ascribed to humans alone to the media and their autonomous forces. These forces re-inscribe into the environment what they seemingly only represent.

Sörlin and Wormbs (2018) distinguish between three types of environing technologies: writing, sensing, and shaping (p. 7). However, it is important to keep in mind that these cultural techniques do not follow each other in a linear fashion, but rather modify each other recursively. Therefore, I suggest modifying the model of Sörlin and Wormbs to foreground that all environing technologies are recursively coupled with each other in such a way that it is impossible to tell which one precedes the others. *Sensing*—to which we may add *sampling* (which includes filtering, photographing, videographing, scanning, etc.)—and *writing* (which also includes counting, classifying, gridding, etc.) are, at the same time, as Karen Barad (2003) argues, "dynamic (re-)configuring(s) of the world" (p. 816), that is, *shaping*. *Writing* is not constrained only to documenting, in one way or the other, what has previously been sampled or sensed. As we will see, writing is also a filter, in the sense of a sampling device, and is always already writing what it samples because it is designed according to some kind of mathematics that prepares the "real" to be symbolized and represented by statistics. A photographic image, which is a sensing device, does not only document microbial information; instead, like all technical media, it introduces the difference between noise and information, and thereby it shapes what it writes and writes what it shapes. The cycle of environing operations (sampling, sensing, writing, shaping) is what creates

"MediaSeas." Each cultural technique (be it filtering by means of a net, or visualizing and classifying by means of a digital imaging system) "is a specific agential practice/intra-action/performance through which specific differences are enacted" (Barad, 2003, p. 816). Differences like the one between map and territory, which, however, turns out to be a principally unstable one. The media history of microbiological oceanography, I hope to show, proceeds through unstable distinctions between medium and message, noise and information, which are enacted through environing media. Although not exclusively specific to the oceans of the High North Atlantic, the methods and media of plankton research that I will discuss in the following, have—by historical contingency—turned the microbial waters of the Northern part of the Atlantic (including the North Sea) between the 50th and the 63rd parallel into an exemplary MediaSea.

"Philosophical Dirt": The Plankton Expedition of 1889

It is rather safe to say that it was the operative linking of filters and microscopy that marked the beginning of microbiological oceanography. The Danish naturalist Otto Friedrich Mueller is usually credited with being the first to use netting to sample seawater that he later examined under a microscope (Taylor, 1980, p. 512). But as early as the middle of the 18th century, the Dutch physician Job Baster on Zeeland was able to identify infusorians as the cause of the sea glow by running North Sea water through blotting paper and placing the remaining filtrate under a powerful microscope.[2] Theories about the causes of marine phosphorescence are at the beginning of the physiological or biological conception of seawater. Christian Gottfried Ehrenberg (1834) already diagnosed "the complete transition from the physical conception [of the sea] to the physiological in a probably impartial, well-credited scholar" (p. 462)[3] on the basis of Johann Kaspar Horner's reports of "filtered luminous water" from the Krusenstern expedition to Brazil in 1802–1804.

Filters are cultural techniques, which are of special interest with regard to theory, because they are material-discursive hinges between the Symbolic and the Real, in the Lacanian sense. We can here draw on Khashayar Razghandi's and Emad Yaghmaei's (2020) "active filter model" (which in turn draws on the theory of cultural techniques), according to which filters are not simply mere objects, but operative systems.

> Our understanding of filters, in this model, requires a similar shift[4] from taking the focus from the filter as an object, article, or device to the filter operations. To think "filter" is to think "filtering." As systems, filters are operative (p. 19).

Insofar as filters are operative systems, they must be viewed as processes through which differences are generated. "Filtering creates Filter domains, building a dynamic ecology" (p. 21). If we link this theory to the conception of media as infrastructures of being, we can go one step further and formulate: Filters are not only environing media; *as environing media* filters create filter ontologies—in our case ontologies of seawater.

The difficulties inherent in distinguishing between medium and content characterized the attempts to determine the cause of marine phosphorescence early on. In 1834, Ehrenberg was able to identify the unicellular organism *Noctiluca scintillans* as the cause of the marine phosphorescence; however, he was not at all sure whether one should make a distinction between the sea water and the cause of the marine phosphorescence. At least, Ehrenberg did not rule out the possibility that there is a transitional stage between seawater and masses of *Noctiluca scintillans*: that is, slime. Seen from the ship, masses of *N. scintillans* "make large sea surfaces [...] appear slimy." Ehrenberg therefore does not want to reject the conjecture outright, cautiously expressed by Alexander von Humboldt in 1826, that "the sea may perhaps be regarded as a gelatinous fluid" (Ehrenberg, 1834, pp. 555, 559). In Humboldt's supposition, one hears an echo of the ideas about the "mucus of the sea" spread among naturalists in the second decade of the nineteenth century. For Bory de Saint-Vincent (1826), the *matière muqueuse* was a "rudimentary means" destined to provide "a milieu for the first operations of organization" (pp. 267–277). Saint-Vincent referred to Geoffroy de Saint-Hilaire, who (according to Saint-Vincent) is said to have written in his work on the *Monstruosités humaines* (1822) about slimy matter that it is the "*substance animalisable par excellence*" (Geoffroy de Saint-Hilaire, quoted in Saint-Vincent, 1826, p. 253). The sea water appears as primordial slime, as protoplasm of all life. As is well known, this theory was taken up again by Jules Michelet in 1861 in his book *La mer* and exaggerated in a fantastic way. Although Saint-Hilaire does not discuss a "*substance animalisable*," but rather a "*substance assimilable*" (Azoulai, 2019, p. 162), Michelet radicalized the theories about the mucus of the sea by turning it into a theory of seawater that sees in everything that can be filtered from the sea only metamorphoses of the medium itself. Already Saint-Vincent had proposed that the infusoria and the salps were nothing more than thickenings of the mucilaginous matter or of the sea water itself (Saint-Vincent, 1826, p. 255). Michelet, however, goes far beyond that. According to Michelet (1861/2006), the jellyfish is nothing other than "gelatinous water" (p. 92) and also the fish, this "glittering, slippery creature with its silver scales [seemed]" to him already as a child "to be identical with the element in which it swam." The fish is "nothing other ... than water itself: animal, organically formed water" (Michelet, 2006, p. 91). According to Michelet's radical media theory, there is no content of the medium at all, but only metamorphoses of the medium. The glitter and slickness of the fish is the glitter and slickness of the sea water. In Michelet's biology of the sea, there are only gradual gradations of the organizedness of matter; what is form is only a slightly more organized version of formlessness. The sea is an elemental medium, full of objects that are object-averse, a sea of abjects.

In 1845, Johannes Müller and his students began to collect "upwelling" on the German archipelago Helgoland with the aid of nets on poles. Müller used the so-called "silk bolting cloth" used by millers to separate different grades of flour (hence the German term *Müllergaze* (Lussenhop, 1974, p. 325)), which was then the finest available netting. Although Müller's studies are of a systematic and morphological nature, he always combined this method with research into the development of the respective species. Müller studied echinoderms such as sea urchins or starfish. In the search for the larvae and developmental forms of these echinoderms, he began to concern himself with what others ignored and what initially only had the status of "admixture" (Lohff, 1992, p. 38). Müller called it "pelagic upwelling" (*pelagischer Auftrieb*), and correspondingly, he termed the method of filtering seawater "pelagic fishing." Systematic morphological biology encounters a massive figure-ground problem as soon as biologists start paying attention to those phenomena of the underwater world that are invisible to the unaided eye. After all, it is not easy to distinguish between signal and noise. What of all those forms and figures which become visible under the microscope is part of the symbolic order of biology—the system—and what is just noise? What belongs to the medium, what belongs to the message? However, "pelagic fishing" meant not only fishing *in* the open sea but also fishing *of* the open sea. The *Müllergaze* does not, like a fishing net, address the organisms which distinguish themselves from the ocean in that they can move of their own accord. Instead, it addresses the free, suspended particles which belong to the milieu. It can be said that Müller was the first marine biologist in an environmental sense because he recognized that there was not just a biology of fish and other aquatic creatures, but a biology of the sea itself.

One of Müller's students was Victor Hensen. In retrospect, Hensen revealed to an audience of biologists in 1906 that Müller had described what had been "overlooked" by the other systematists as "philosophical dirt" (Hensen, 1906, p. 365; Kölmel, 1986, p. 66; Lohff, 1992, p. 38). I cannot think of a more beautiful synonym for the Lacanian Real.[5] Philosophy has to do with the logos, marine science with what Georges Bataille called the Heterological, the waste, the Formless. However, Hensen saw something different in Müller's "dirt" than Müller himself did—namely, it was not a noise from which one had to filter out the object relevant to the completion of the system (larvae of echinoderms or cyclotomes), but rather it was the nutrient basis of the fish that lived in the sea. The pelagic upwelling had to be the primal nutrition of all larger marine animals. Hensen's research was motivated not by a zoological interest but by an economic one: What was to be researched was not the individual creature, its morphology and development, but rather the total production of organic matter by the sea, against the background of a possible overfishing of the seas. Therefore, the quantitative recording of the primal nutrition of marine animals had to be the key to studying the overall metabolism of the ocean.[6]

In 1864, Hensen was appointed to the Chair of Physiology at the University of Kiel. In 1870, at his instigation, the Prussian State Parliament, of which Hensen was a member, initiated the founding of the Commission for the Scientific

Investigation of the German Seas. In 1887, Victor Hensen named the pelagic upwelling plankton. Two years later, he launched the plankton expedition with the support of the Royal Prussian Academy of Sciences and the Kaiser himself.

The aim of the plankton expedition was to determine how the plankton is distributed in the sea, as well as whether there was a connection between the occurrence of plankton and biodiversity, solar radiation, and the climate zones. The plankton expedition was the first oceanographic expedition to conduct exclusively quantitative research. The most astonishing result of the expedition was that the seas in the southern, and therefore warmer, climate zones and in particular the Sargasso Sea are extremely poor in plankton, while the cold High North Atlantic is extremely plankton-rich.

Hensen and his colleagues presented the results of their expedition in a document that is both a map and a statistical diagram (Figure 2.1). The charted travel route serves as abscissa (as x-axis), on which, at the respective points at which catches were made, the quantity of those catches is plotted as the ordinate (as y-axis). You can see the comparatively gigantic deflections of the ordinate within the High North Atlantic, which at about 60° N 28° W even exceeds the frame of the map. But apart from the content of the map, this document also makes a statement about the *medianature* of the sea. The map makes the statement that the *National*, the ship of the plankton expedition, sailed across the Atlantic and at the same time across numbers (which indicate the size of the plankton catches). The sea is no longer a vast geographical space; but rather, is a hybrid of geography and data—a sea of data.

The main medium through which the plankton expedition produced this sea of data was the plankton net devised by Hensen, and the main cultural techniques were filtration and counting (Hensen, 1893, p. 67). From now on the net, a material thing, is always already part of a symbolic order; a symbolic order, however, that is no longer subordinate to the logos but to statistics. Dirt is not articulated through language but through counting and sorting. To be of any value, the samples had to be representative of the area studied. To do this, Hensen had to develop a technique to filter out all the plankton from a precisely defined volume of water in order to obtain a basis for calculating the extrapolation. The solution to this problem was the "vertical plankton net." It is lowered to a certain depth (200 m) and then hauled up again vertically. Its construction principle guarantees that the entire water column located above the net opening is completely filtered. The volume of the filtered water column then results from the product of the height of the water column determined by depth measurement and the surface area of the net opening. Representative subsamples were then taken from the catches, which were then classified, sorted, and counted under the counting microscope designed by Hensen himself.[7]

Today it is a "truism in oceanography that all processes operating at finer scale than the resolution of the sampling grid are suppressed or ignored, implicitly if not explicitly, by the observer" (Herman and Platt, 1908, p. 204). Hensen's work was not an exception to this rule. Hence, what Hensen filtered was not plankton as such, but so-called "net-plankton," a term introduced by Hans Lohmann, a fellow

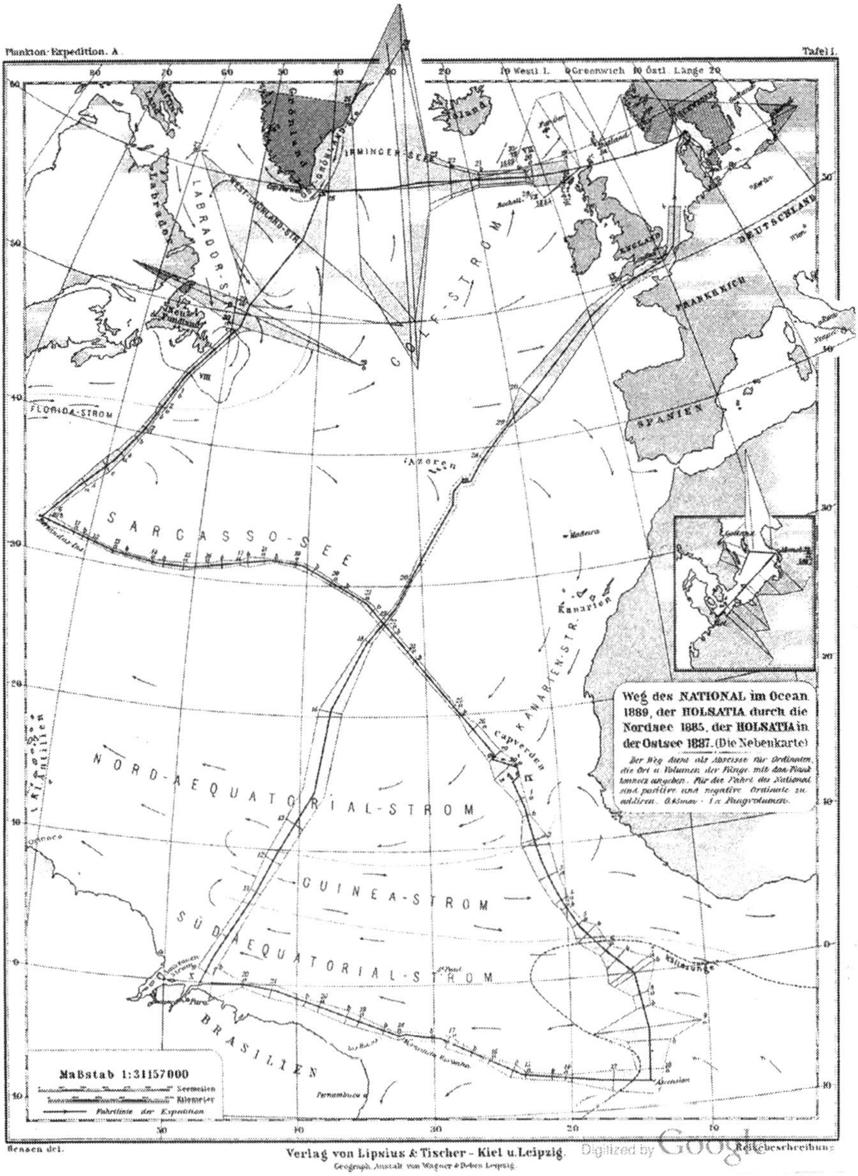

Figure 2.1 The journey of the *National* in the Atlantic Ocean 1889

Note: The translation of the text in the box reads: "The route serves as axis of abscissae for the ordinates which indicate place and volume of the catches with the plankton net. For the journey of the *National,* the positive and the negative ordinate has to be added."

Source: Otto Krümmel: Reisebeschreibung der Plankton-Expedition, Kiel/Leipzig 1892 (= Ergebnisse der Plankton-Expedition der Humboldt-Stifung, Bd. IA), plate I.

participant in the plankton expedition, to describe the plankton that can be filtered using a net. However, there is no plankton as such; all plankton is a "*medianature.*" The *Müllergaze* Hensen used had a mesh size of 0.0002368 square millimeters (Hensen, 1893, p. 73). Everything below the filtration threshold of *Müllergaze* is commonly named nanoplankton, a term introduced by Lohmann in 1909. Nanoplankton is everything which cannot be filtered from seawater using a net. If net-caught plankton defines plankton as such (as was the case for Hensen), everything that is not plankton is seawater. "Filtering creates Filter Domains." (Razghandi and Yaghmaei, 2020, p. 21). However, if seawater is nothing but non-filterable plankton, then there is no seawater, only different orders of magnitude of plankton, or, in other words: only different cultural techniques of filtration.

Flicker Films of the High North Atlantic: The CPR Survey

The cultural technique of counting became the guiding paradigm of plankton research for the next decades—at least in the North and High North Atlantic. In the 1920s, the British marine biologist Alister Hardy conceived the Continuous Plankton Recorder (CPR) survey as a means of mapping near-surface plankton in space and time, interpreting the changing fortunes of the fisheries and relating plankton changes to hydrometeorology and climatic change (Reid et al., 2003; Gamble and Hunt, 1992). The medium of this survey, however, which started in 1931 and has grown to become the most extensive long-term survey of marine organisms in the world, was not the plankton net anymore, but a device invented by Hardy in the 1920s: the Continuous Plankton Recorder. However, the central term of the whole enterprise—*recording*—already points to a fundamental ontological disruption of the concept of representation, which is due to the interpretation of the cultural technique of "sampling" as "recording." That "sampling" has now been understood as a method of recording, carries with it the obvious reason that the functional principle of CPR has been borrowed from a film camera. By "recording" the plankton, the CPR produces that ontological disruption or category confusion that it transforms the plankton into an image of itself. Sampling and recording, the object and its image, fall into one.

The CPR is towed by Ships of Opportunity (SOOPs)[8] at a known speed (usually 15–20 knots); this allows seawater to enter the apparatus through a 1.62 cm^2 entrance aperture; the plankton is filtered through a continuous tape of filtering silk unwound from a spool, covered by a second tape of covering silk, and spooled onto a second spool in a storage tank containing formalin to preserve the biological specimens that are collected (Richardson et al., 2006). The sampling grid of the CPR survey is constructed from the ship routes from east to west and vice versa on the one hand (constituting the irregular parallels of the grid) and the sections on the tape of the recorder on the other hand (constituting the meridians of the grid). Here, 1 cm on the gauze tape corresponds to 1 nautical mile. In the laboratory, the gauze roll is cut into sections equivalent to 10 nautical miles (18.5 km) or approximately 3 cubic meters of filtered seawater. The green coloration of the sieve is determined to obtain an index of chlorophyll concentration.

Phytoplankton taxa and smaller (< 2 mm) and larger (> 2 mm) zooplankton species per section are then identified and counted. This datafication of the North Atlantic eventually resulted in the publication of the first *CPR Atlas of the North Atlantic* in 1973 and the second in 2004, based on 155,669 CPR samples taken monthly along specific routes in the North Atlantic between 1958 and 1999. It shows the geographical distribution of 240 species (Continuous Plankton Recorder Team, 2004). A grid was established, ranging from 70°W to 20°E and from 30°N to 70°N. The size of the geographic grid cell for spatial regularization was set at 100 × 100 nautical miles, which is a compromise between the number of missing grid cells and resolution (Beaugrand, 2004). The North Atlantic is transformed into a raster screen, with a resolution of 41 pixels in length and 26 pixels in width. Each pixel is assigned a discrete gray value (it is a black and white screen) that indicates the abundance of a particular plankton species within a bimonthly time interval. If the abundance of the particular species exceeds or falls below a certain threshold, a darker or lighter gray value is assigned to the pixel.

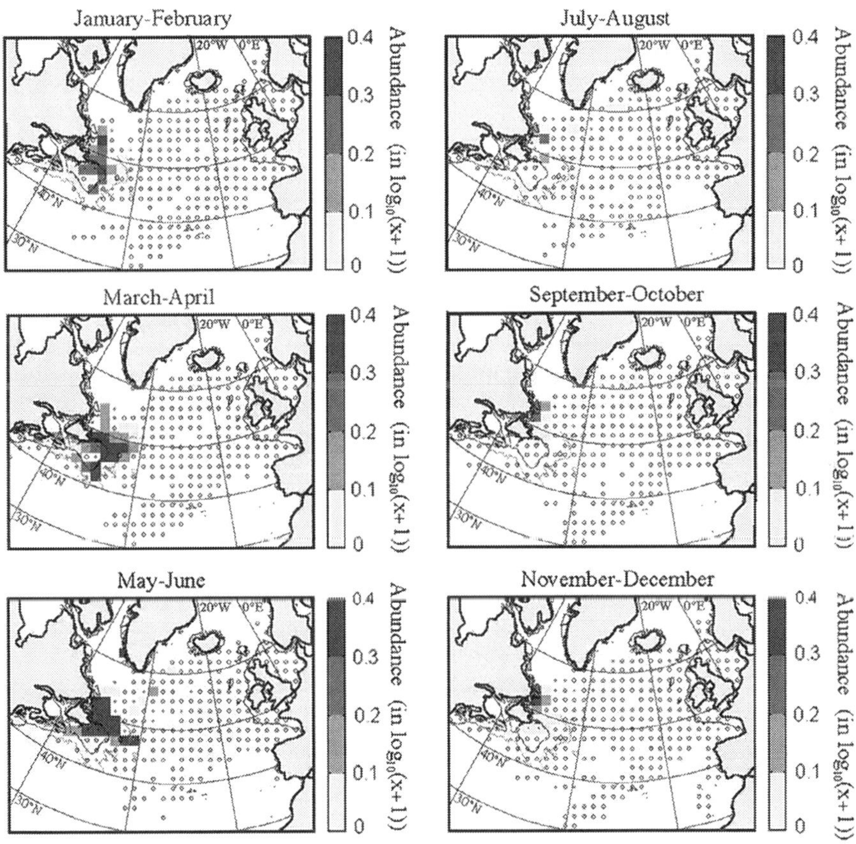

Figure 2.2 The *Calanus glacialis* "movie." Seasonal changes in geographical distribution (nighttime data)

Thus, the North Atlantic was transformed into a screen on which 240 flicker films were played, each consisting of two-dimensional frames made up of gray values that follow one another at a frame rate of two months—"films" that show the statistical distribution of 240 plankton species at a given image resolution of 41 x 26 pixels (Figure 2.2).[9]

In 1999, the CPR survey was adopted as an integral part of the Global Ocean Observing System (GOOS) (Rio Fernandez et al., 2019, p. 335), which aims at an "integrated digital ocean." According to Jessica Lehmann, the GOOS is "responsible for the creation of a new concept of the world ocean, or perhaps more accurately a new world ocean: an ocean of data, a digital doppelganger for the wet and wild ocean out there, an ocean made informational" (Lehmann, 2016, p. 113). The recent media history of plankton research is characterized by exactly this paradigm shift: data extraction from images replaces data extraction from plankton samples.

From Net to Operational Image

Even if Johannes Müller's pelagic fishery, Victor Hensen's wondrous hybrid map, and the 2004 plankton atlas are worlds apart, one thing had not changed in the plankton discourse network between 1820 and 1980: The "body and mind ruining" work of counting (as Ernst Haeckel would debunk it). As late as the 1970s, "direct microscopic enumeration" was "the only approach giving information detailed enough for many ecological studies" (Ortner et al., 1979, p. 50). In 1978, marine biologists at the National Oceanic and Atmospheric Administration in Miami, along with Harold E. Edgerton, a pioneer of ultra-high-speed photography at MIT, introduced a technique that used high-speed silhouette photography to evaluate living zooplankton samples at sea (Ortner et al., 1979, pp. 50–51). Seawater samples were placed in a darkroom in a plastic tray directly on the film, and the film was exposed with a flash, the duration of which was 3×10^{-6} seconds. The validity of the method of quantitative plankton research developed by Victor Hensen was based on the fact that the net content corresponded to the precisely known volume of a filtered water column. In the silhouette photography process, the photonegative now took the place of the mesh: the dimension of the negative corresponded "to a known volume of ocean" (443 m^3 of seawater filtered through a 0.333 mm mesh) (Ortner et al., 1979, p. 51). However, once again the degree of resolution determines what is figure and what is ground: the resolution of the photographic material must be at least equal to the resolution of the filtering tissue so that the photograph can be used to quantitatively characterize the plankton of a given area. Therefore, what constitutes the granular background of the magnified silhouette of a copepod or a pteropod (a coarse-grained noise, so-called "marine snow") is a mixture of seawater material and photographic material. Where the granularity of the photographic material is of the same order as the granularity of the "admixture," it becomes undecidable what is medium and what is image, what is granulation of the photographic material and what is granulation of the seawater. Whatever enters the symbolic order as plankton, its ontology is determined by a ratio to this Real, conditioned by media technology, in which the granules of the seawater material are indistinguishable from the granules of the photographic material.

Since the 1980s, methods have been developed to fully automate the enumeration of zooplankton by means of direct video recording techniques and real-time computerized image analysis procedures. In the process, the net and the camera (which in 1978 were still separated in time and space—the net in the sea, the camera on board the ship) moved closer and closer together until the net was finally completely integrated into the imaging process. In the beginning, cameras were placed directly in the cod-end of the plankton net. The inventor of the silhouette photography method, Peter Ortner, developed a field application of the laboratory-based method in 1981 by placing a 35 mm camera with a high-capacity film magazine in front of the cod-end of a conical plankton net attached to a rigid frame (Wiebe and Benfield, 2003, pp. 60–61). Ten years later, a television camera replaced still cameras for the first time. In July 1991, a research group at the Kiel Institute of Oceanography (a member of the EUROMAR project) tested the Ichthyoplankton Recorder (IR) in the Bornholm Basin of the Baltic Sea, in which a "high-resolution" video camera had been integrated into the cod-end of a standard German plankton sampler. Separate interlaced video fields were transmitted via a line cable aboard the research vessel *Alkor*, where they were recorded at 50 Hz on S-VHS. The 50 Hz sampling rate resulted in a horizontal spatial resolution of 5.14 cm at a towing speed of 5 knots (i.e., an image is obtained every 5.14 cm during towing) (Lenz et al., 1995, p. 62; Welsch et al., 1991, p. 2). However, people on board the *Alkor* did not watch video. Instead, the video images were analyzed frame by frame (each frame was equivalent to 600 ml of seawater). However, the 3000 images per minute provided by the video system could no longer be evaluated manually. For copepods, therefore, only every 30th frame was evaluated; for counting fish eggs, the intervening frames were run in slow motion (Welsch et al., 1991, p. 3).

Eventually, the video image replaced the plankton net entirely. The Video Plankton Recorder, introduced in 1992, was a plankton sampling device that could automatically identify and count phytoplankton and zooplankton, mapping their abundance and distribution in real time and at high resolution (Benfield et al., 2007, pp. 173–174). It consisted of four CCD cameras, each having a different magnification, a stroboscope, a plankton recorder box, and an image processing system (Davis et al., 1992). But was it so easy to completely replace the object with its image? Was it that easy to completely surrender scientific truth to the virtuality of the image? Could plankton research completely abandon its referents without reservation to become a pure science of signifiers? The answer is no. "To ground truth the video image" (Davis et al., 1992, p. 72),[10] directly behind the volume of water captured by the camera with the highest magnification is mounted an LHPR box[11] that collects the plankton material captured by the cameras on a Nitex gauze strip. Since the data from the LHPR, which counts the frames of the gauze, is read along with the video time code, direct comparisons between the video images and the actual specimens collected on the gauze are possible. The signifiers are still chained to their referents, even if only by synchronized counters.

In the meantime, numerous so-called "in situ imaging systems" have been added, with the help of which quantitative images of meso- and also nano-plankton can be "collected." The gap between the cultural techniques of filtering and imaging is closing. Imaging ceases to visualize previously filtered plankton samples through a tedious and complicated process of identifying, counting, averaging, and mapping, but coincides with the techniques of filtering, identifying, counting.

Widely used today, the Underwater Vision Profiler 5 (UVP5) (first tested in 2008) can be used to obtain digital images of particles and zooplankton >100 microns. The images are taken by means of a flash light illuminating an area whose volume is exactly 1 liter of seawater per image (Picheral et al., 2010, p. 465). The discourse of "in situ imaging systems" typically no longer distinguishes between the "collecting" of images and the "collecting" of plankton. The seawater is identified with its visualized and digitized information. Essential data about the plankton captured by the camera is obtained by operationalizing the digitality of the images. Thanks to a size calibration, the UVP5 can convert the pixel area occupied by a particle into millimeters (Picheral et al., 2010, p. 466). The alternative to in situ camera systems are scanners that digitize traditionally net-fished plankton samples without the detour of photography. The most widely used is ZooScan, a scanner introduced in 2004 that has a watertight chamber instead of a flatbed for scanning images, into which zooplankton samples can be poured and digitized at high resolution (Grosjean et al., 2004).[12] The contents of these digital images of plankton samples are then identified, counted, and measured by various classification algorithms or trained machine-learning neural networks.[13] The effectiveness of imaging systems thus depends more and more on the classification software used; for example, ZooScan or the UVP5 work with a software called Plankton Identifier, which allows automatic identification of object images previously extracted with another piece of software called ZooProcess (Gorsky et al., 2010). The implemented Plankton Identifier algorithm (Random Forest) learns to classify unknown images using decision trees and to associate the features extracted from the images with metadata. The hoped-for result is that "useful information" is extracted from "the vast numbers of images" produced by in situ imaging systems like UVP5 or scanners like ZooScan, which are not coincidentally called "sampling approaches" (Benfield et al., 2015, p. 177). The difference between sampling vast numbers of images and sampling vast numbers of plankters, the single organisms of plankton, is increasingly melting away. Plankton image classification, however, presents machine vision with a major challenge: not only is plankton morphologically very heterogeneous (there are metamorphoses between the larva and the adult animal in which the animal can assume extremely different forms), it also inhabits "a medium that also contains a variety of nonliving targets such as marine snow, sediment particles, and bubbles" (Benfield et al., 2015, p. 178). The metaphor of snow is noteworthy: as is well known, the noise of the cathode ray tubes of early low-definition TV was also referred to as "snow." Snow, a milieu-specific feature of the High North Atlantic, forms a continuum that connects seawater and image screen. Snow, which in 2007 was attributed to the medium as background noise, is now also identified as an Essential Ocean Variable (EOV), like the biomass and diversity of phytoplankton and zooplankton, because the particle aggregates called "marine snow" are very often colonized by microorganisms. These particle aggregates can sink rapidly and form one of the major fluxes of matter into the deep ocean and are an important vector of nutrient distribution in the ocean (Lombard et al., 2019). It remains a fact: the media history of microbiological oceanography from Johannes Müller to the OceanObs'19 conference proceeds through unstable distinctions between medium and message. The distinction between non-sense and sense, between what one must ignore and what one must perceive, is always only temporarily stable. What was snow a moment ago is an EOV a little later. The distinction between noise and information is not an ontological one, but a technically contingent one. What was noise yesterday is information today. It is still true today that plankton as such does not exist, but only the "effective range" of the media systems used in each case.[14]

The target size for quantitative observations of ZooScan is between 240 und 5000 μm ESD[15] (Lombard et al., 2019, p. 7.). Flow imaging microscopy can be used for visualization of smaller target sizes. For target sizes down to 0.2 μm, which are needed for example to distinguish cyanobacteria of different sizes and for counting them, flow cytometers are used (Helmreich, 2009, p. 181). The water sample is streamed through a laser beam, and since every cell scatters the light in a different way according to its size, bacteria and viruses can be separated and counted.

In 1988, MIT's Sallie Chisholm discovered what is arguably the world's most abundant photosynthetic marine bacterium, Prochlorococcus, in the Sargasso Sea by means of a flow cytometer (Chisholm, 2012). Most recently, with the discovery of Prochlorococcus, the ontology of seawater is of a genomic nature, which is evident from amino acid sequences in databases. Other cultural techniques have been added to the filtration which constitute the media apriori of seawater's medianature in the 21st century: "Collecting, filtering, scanning, preserving, and distributing water samples represent techniques that ready the sea to be uploaded … to databases" (Helmreich, 2009, p. 177). For Hensen, in 1890, only what could be filtered by a plankton net was plankton. Now, what is considered is only what can be stored in a database. "When microbial DNA fragments stand in for seawater," Melody Jue (2020) writes, "the logic of digitality appears commensurate with seawater" (p. 134).

However, the logic of digitality already applied to digital imaging techniques like the VPR, ZooScan, and the UVP5.

However, to even begin to master the millions of images "collected" by oceanographic research worldwide each year, an infrastructure is needed to link ocean datafication projects as well as to generate a geography of plankton from these projects. EcoTaxa is an online database that stores samples (images) collected and automatically identified, for example, with UVP5 and ZooScan to allow scientists to classify large volumes of images (Lombard et al., 2019, p. 14). The accuracy of the algorithms used was 75% in 2014; human taxonomists increase the recognition level to over 85%. To ensure that the remaining 15% are not simply ignored as noise, the symbolic order of marine biology in EcoTaxa also recognizes a phylum called "artifact," whose classes are named "badfocus," "bubble," "puff," "tuff," "turbid," or "othertocheck." The paradigm shift has occurred—almost silently and almost unnoticed: the problem now, mind you, is to classify vast quantities of *images*, no longer vast quantities of *plankters*—protozoa, annelids, molluscs, arthropods, phoronida, and so on. The object of oceanographic media infrastructures is no longer the oceans, but their digital doppelgangers.

Notes

1 See also Peters and Wickberg (2022, p. 676): "Media can be profitably understood as environing and residual, not only semiotic and emergent."

2 See Baster (1762, pp. 31–32): "the small luminous sparks that we see, especially in the summer … in innumerable abundance on the surface of the moving and evaporating sea water, (are) very small animals, which, as is well known, are only visible under the microscope. To collect them, we take a hemina (= 0.274 l) full of sea water, in which these bright sparks are seen at night, the water is filtered until it is sufficient to fill a spoon: a drop of it is placed on a glass plate, and the microscope, at great magnification, will show a large number of the animals mentioned, of which I have distinguished three species." (My translation.)

3 Horner, who was actually an astronomer, was inspired to study seawater by the research of Georg Heinrich von Langsdorff, who also took part in the Krusenstern expedition.

4 Similar to the shift made by the theory of cultural techniques from objects to chains of operations.

5 Lacan has given more than one (at least three) different definitions of the Real ("le réel"). Here I refer to the definition he gave in the preface to the German edition of his selected Writings: The Real is "what does not cease not to be written down." See Lacan (1975, p. 14).

6 On Hensen and the Plankton Expedition see Finstad, 2017.

7 The stage of the counting microscope can only be moved in the direction of rectangular coordinates. A glass plate ruled by a diamond is placed on the rectangular stage (size 182 sq cm). The lines are so close together that even at 200x magnification two lines are still visible in the eyepiece. This ensures that all individuals on the stage are counted and that no individual is counted twice. See Hensen (1893, pp. 145–147).

8 SOOPs are ordinary merchant ships that voluntarily act as platforms for sensors carried or towed by them.

9 However, the raster images published in the Plankton Atlas do not show the "movie sequences," but images of the average spatial distribution obtained from the "movie sequences."
10 The term "ground truth" is used here in a similar way as it is used in geography where it "designates the information provided by direct observation—usually at the level of the literal ground, the surface of the Earth—in relation to maps, models, and remote sensing technologies." See Gil-Fournier and Parikka (2021, p. 1254). The use of the term in the citation above therefore is quite ironical since in oceanography there is no ground in the literal sense.
11 The Longhurst Hardy Plankton Recorder (LHPR) is the most widely used plankton sampler. See Harris et al. (2000, p. 60).
12 The samples are digitized at a resolution of 2400 dpi, and the resulting operational images have a size of 17,500 x 7000 pixels. The pixel size is 10.58 mm, which means that with a plankton net mesh size of 200 mm used, one net mesh corresponds to 18.18 pixels.
13 From the "images" created with ZooScan, the »objects« were first extracted and then divided into two training sets. These training sets were then used to test a number of classification algorithms: "linear, quadratic, mixture, and flexible discriminant analysis, k-nearest neighbors; learning vector quantization, tree and recursive partitioning methods, including ensembles of bootstrapped tree, such as bagging and random forest, support vector machine, feed-forward, single hidden layer neural network." See Grosjean et al. (2004, p. 522). Truly, marine science has become computer science.
14 "Small particles are often too small to be imaged efficiently (too few pixels, signal below threshold or near noise level) while larger organisms are too scarce to be sampled quantitatively (volume analyzed is too small), or too large to pass by the tubing of some devices, resulting in a narrow size range compared with the theoretical one. This 'effective' range is often missing from technical documentation ..." See Lombard et al. (2019, p. 6).
15 ESD = equivalent spherical diameter.

References

Angerer, M.L. (2020). Affective milieus: Intensive couplings, technical sentience, and a nonconscious in-between. In B. Bösel and S. Wiemer (Eds.), *Affective transformations: Politics-algorithms-media* (pp. 87–101). Meson Press. doi:10.25969/mediarep/15024.

Azoulai, J. (2019). Métamorphoses et biologie marine chez Michelet et Richepin. In J. Azoulai, A. Fayolle, and G. Séginger (Eds.), *Les métamorphoses entre fiction et notion. Littératures et sciences (XVIe XXIe siècles)* (pp. 159–174). LISAA éditeur.

Barad, K. (2003). Posthumanist performativity: Toward an understanding of how matter comes to matter. *Signs: Journal of Women in Culture and Society*, 28(3), 801–831. doi:10.1086/345321.

Baster, J. (1762). *Opuscula subseciva, observationes miscellaneas de animalculis et plantis quibusdam marinis, eorumque ovariis et seminibus continentia*. Johannes Bosch Harlem.

Beaugrand, G. (2004). Continuous plankton records: Plankton atlas of the North Atlantic (1958–1999). I. Introduction and methodology. *Marine Ecology Progress Series*, (n.v.)(Suppl. 1), 3–10.

Benfield, M.C., Grosjean, P., Culverhouse P.F., Irigoien, X., Sieracki, M.E., Lopez-Urrutia, A.*et al.* (2015). RAPID: Research on automated plankton identification. *Oceanography*, 20(2), 172–187. doi:10.5670/oceanog.2007.63.

Chisholm, S.W. (2012). Unveiling prochlorococcus: The life and times of the ocean's smallest photosynthetic cell. In R. Kolter and S. Maloy (Eds.), *Microbes and evolution: The world that Darwin never saw* (pp. 165–171). ASM Press. doi:10.1128/9781555818470.ch23.

Continuous Plankton Recorder Team (2004). Continuous plankton records: Plankton atlas of the North Atlantic (1958–1999). II. Biogeographical Charts. *Marine Ecology Progress Series*, (n.v.)(Suppl. 2), 11–75.

Davis, C.S., Gallager, S.M., Berman, M.S., Haury, L.R., and Strickler, J.R. (1992). The Video Plankton Recorder (VPR): Design and initial results. *Archiv für Hydrobiologie. Beiheft Ergebnisse der Limnologie*, 36, 67–81.

Ehrenberg, C.G. (1834). Das Leuchten des Meeres. Neue Beobachtungen nebst Übersicht der Hauptmomente der geschichtlichen Entwicklung dieses merkwürdigen Phänomens. *Abhandlungen der physikalisch-mathematischen Klasse der Königlichen Akademie der Wissenschaften zu Berlin. Aus dem Jahre 1834*, 411–577.

Finstad, T. (2017). A sea of microbes: Changing representations of oceans and life on earth. Typescript.

Gamble, J.C., and Hunt, H.H. (1992). The Continuous Plankton Recorder Survey: A Long-Term, Basin-Scale Oceanic Time Series. In NASA, Godard Space Flight Center (Ed.), *Proceedings of the Ocean Climate Data Workshop* (pp. 277–293).

Gil-Fournier, A., and Parikka J. (2021). Ground truth to fake geographies: Machine vision and learning in visual practices. *AI & Society*, 36, 1253–1262. doi:10.1007/s00146-020-01062-3.

Gorsky, G., Ohman, M.D., Picheral, M., Gasparini, S., Stemmann, L., Romagnan, J. B., Cawood, A., Pesant, S., García-Comas, C., and Prejger, F. (2010). Digital zooplankton image analysis using the ZooScan integrated system. *Journal of Plankton Research*, 32(3), 285–303. doi:10.1093/plankt/fbp124.

Grosjean, P., Picheral, M., Warembourg, C., and Gorsky, G. (2004). Enumeration, measurement, and identification of net zooplankton samples using the ZOOSCAN digital imaging system. *ICES Journal of Marine Science*, 61(4), 518–525. doi:10.1016/j.icesjms.2004.03.012.

Harris, R., Wiebe, P., Lenz, J., Skjoldal, H.R., and Huntley, M. (Eds.) (2000). *ICES zooplankton methodology manual*. Academic Press. doi:10.1016/B978-0-12-327645-2.X5000-2.

Helmreich, S. (2009). *Alien ocean: Anthropological voyages in microbial seas*. University of California Press. doi:10.2307/j.ctv1 1hpsjh.

Helmreich, S. (2011). Nature/Culture/Seawater. *American Anthropologist*, 113(1), 132–144. doi:10.1111/j.1548-1433.2010.01311.x.

Hensen, V. (1893). Methodik der Untersuchungen. In V. Hensen (Ed.), *Ergebnisse der Plankton-Expedition der Humboldt-Stiftung* (Vol. 1B) (pp. 1–200). Verlag von Lipsius & Tischer.

Hensen, V. (1906). Die Biologie des Meeres. Rede am Stiftungsfest des naturwissenschaftlichen Vereins in Kiel. *Archiv für Hydrobiologie und Planktonkunde*, 1, 360–377.

Herman, A., and Platt, T. (1980). Meso-scale Spatial Distribution of Plankton: Coevolution of Concepts and Instrumentation. In M. Sears and D. Merriman (Eds.), *Oceanography: The Past. Proceedings of the Third International Congress on the History of Oceanography, held September 22–26, 1980 at the Woods Hole Oceanographic Institution, Woods Hole, Massachusetts, USA on the occasion of the Fiftieth*

Anniversary of the founding of the Institution (pp. 204–225). Springer New York. doi:10.1007/978-1-4613-8090-0_21.

Hessler, S. (2019). *Prospecting ocean.* MIT Press. doi:10.4000/critiquedart.61616.

Jue, M. (2020). *Wild blue media.* Duke University Press. doi:10.1515/9781478007548.

Kölmel, R. (1986). Victor Hensen als Meeresforscher. *Biologie in unserer Zeit*, 16(3), 65–70.

Lacan, J. (1975). *Schriften* II (N. Haas, Ed.). Walter-Verlag (Originally published in 1966).

Lehmann, J. (2016). A sea of potential: The politics of global ocean interventions. *Political Geography*, 55(3), 113–123. https://doi.org/10.1016/j.polgeo.2016.09.006.

Lenz, J., Schnack, D., Petersen, D., Kreikemeier, J., Hermann, B., Mees, S., and Wieland, K. (1995). The Ichthyoplankton Recorder: A video recording system for in situ studies of small-scale plankton distribution patterns. *ICES Journal of Marine Science*, 52(3–4),409–417.

Lidström, S., Wickberg, A., and Gärdebo, J. (2022). Ocean environing media. Datafication of the deep sea. In A. Wickberg and J. Gärdebo (Eds.), *Environing media* (pp. 114–133). Routledge. doi:10.4324/9781003282891-10.

Lohff, B. (1992). Die Entdeckung der Welt des Planktons. *Historisch-Meereskundliches Jahrbuch*, 1, 35–44.

Lombard, F., Boss, E., Waite, A.M., Vogt, M., Uitz, J., Stemmann, L.*et al.* (2019). Globally consistent quantitative observations of planktonic ecosystems. *Frontiers in Marine Science*, 6(196), 1–21. doi:10.3389/fmars.2019.00196.

Lussenhop, J. (1974). Victor Hensen and the development of sampling methods in biology. *Journal of the History of Biology*, 7(2), 319–337. doi:10.1007/BF00351208.

Michelet, J. (2006). *Das Meer* (R. Wintermeyer, Ed.). Campus (Original work published 1861).

Ortner, P.B., Cummings, S.R., Aftring, R.P., and Edgerton, H.E. (1979). Silhouette photography of oceanic zooplankton. *Nature*, 277, 50–51. doi:10.1038/277050a0.

Parikka, J. (2012). New materialism as media theory: Medianatures and dirty matter. *Communication and Critical/Cultural Studies*, 9(1), 95–100. doi:10.1080/14791420.2011.626252.

Picheral, M., Guidi, L., Stemmann, L., Karl, D.M., Iddaoud, G., and Gorsky, G. (2010). The Underwater Vision Profiler 5: An advanced instrument for high spatial resolution studies of particle size spectra and zooplankton. *Limnology and Oceanogryphy: Methods*, 8(9), 462–473. doi:10.4319/lom.2010.8.462.

Peters, J.D. (2015). *The marvelous clouds. Toward a philosophy of elemental media.* University of Chicago Press.

Peters, J.D., and Wickberg, A. (2022). Media: The case of Spain and new Spain. *Critical Inquiry*, 48(4), 676–696.

Razghandi, K., and Yaghmaei, E. (2020). Rethinking filter: An interdisciplinary inquiry into typology and concept of filter, towards an active filter model. *Sustainability* 12(18), 7284. doi:10.3390/su12187284.

Richardson, A.J., Walne, A.W., John, A.W.G., Jonas, T.D., Lindley J.A., Sims, D.W., Stevens, D., and Witt, M. (2006). Using continuous plankton recorder data. *Progress in Oceanography* 68(1), 27–74. doi:10.1016/j.pocean.2005.09.011.

Rio Fernandez, J. del, Toma, D.M., Martinez, E., Jirka, S. and O'Reily, T. (2019). From sensor to user – Interoperability of sensors and data systems. In E. Delory and J. Pearlman (Eds.), *Challenges and innovations in ocean in situ sensors: Measuring inner ocean processes and health in the digital age* (pp. 289–303), Elsevier. doi:10.1016/B978-0-12-809886-8.00006-5.

Reid, P.C., Colebrook, J.M., Matthews, J.B.L., Aiken, J., and Continuous PlanktonRecorder Team (2003). The Continuous Plankton Recorder: Concepts and history, from Plankton Indicator to undulating recorders. *Progress in Oceanography*, *58*(2–4),117–173. doi:10.1016/j.pocean.2003.08.002.

Saint-Vincent, B. de (1826). Matière. In *Dictionnaire classique d'histoire naturelle* (1st ed., pp. 248–281).

Sörlin, S., and Wormbs, N. (2018). Environing technologies: A theory of making environment. *History and Technology*, 34(2), 101–125. doi:10.1080/07341512.2018.1548066.

Taylor, F.J.R. (1980). Phytoplankton Ecology before 1900: Supplementary Notes to the "Depths of the Ocean." In M. Sears and D. Merriman (Eds.), *Oceanography: The Past. Proceedings of the Third International Congress on the History of Oceanography, held September 22–26, 1980 at the Woods Hole Oceanographic Institution, Woods Hole, Massachusetts, USA on the occasion of the Fiftieth Anniversary of the founding of the Institution* (pp. 509–521). Springer New York. doi:10.1007/978-1-4613-8090-0_47.

Welsch, W., Barthel, K.G., Fröse, R., Hermann, B., Lenz, J. Mees, S., Schnack, D., and Waller, U. (1991). A high-speed video recording system for in-situ studies on small-scale distribution of zooplankton and ichthyoplankton; preliminary results on the distribution of plankton in the Born-Holm Basin (central Baltic). Paper (L:81) presented at the Council Meeting of the International Council for the Exploration of the Sea (ICES). https://www.researchgate.net/publication/29723904_High-speed_video_recording_system_for_in-situ_studies_on_small-scale_distribution_of_zooplankton_and_ichthyoplankton_preliminary_results_on_the_distribution_of_plankton_in_the_Bornholm_Basin_Central_B.

Wiebe, P.H., and Benfield, M.C. (2003). From the Hensen net toward four-dimensional biological oceanography. *Progress in Oceanography*, 56, 7–136. doi:10.1016/S0079-6611(02)140–144.

3 Moving the Green Economy to the Blue Sea

Documents as Sites of Mediation

Marie Stilling and Kristin Asdal

Introduction

What image comes to mind when you think of economic activity in the High North Atlantic? Perhaps the oil rig, the concrete platform resting on pillars above the surface of the ocean with far-reaching cranes and pipes stretching towards the sky? This would be likely since in the past 50 years the High North Atlantic has become a site for oil and gas extraction, with the oil rig the image par excellence of its presence in the High North Atlantic (see Couling, this volume). But now, as the carbon era seems to be coming to an end, a blue bioeconomy—based on biological and renewable resources—is set to take its place. Seaweed cultivation is one industry within this envisioned blue bioeconomy. Although well-established in Southeast Asia, seaweed cultivation is a new industry to these seas, and one that carries hopes of being particularly good for both the economy and the environment.

Seaweed is renowned for its high uptake of carbon and nutrients. It requires neither fertilizers nor freshwater, and where it grows fish can find areas for spawning and rearing fry (Norderhaug et al., 2020). The High North Atlantic, with its strong currents, stable temperatures, and nutrient-rich waters, appears to be particularly fertile for cultivating seaweed (Broch et al., 2019). Now imagine *that*: large, green fields of swaying seaweed taking the place of the oil rigs in the High North Atlantic as well as in the economy as a whole.

Green growth in the blue sea. That is indeed an alluring vision. At present it *is* just a vision.[1] Yet, multiple research and development projects are currently experimenting with cultivating seaweed offshore,[2] albeit that none have so far been able to establish economically viable seaweed cultivation in the High North Atlantic (North Sea Farmers, n.d.; SINTEF, 2024; ULTFARMS, n.d.).

The ambitions for this economy are high, particularly in Norway, which is this chapter's point of departure. One estimate has stipulated that a Norwegian seaweed cultivation industry can produce as much as 20 million tonnes of seaweed biomass annually with a sales value of 40 billion Norwegian kroner—roughly 5.3 billion euro[3]—by 2050 (DKNVS and NTVA, 2012). That is a huge estimate. In comparison, the total global production of cultivated seaweed in 2012 amounted to 23.6 million tonnes (Ferdouse et al.,

DOI: 10.4324/9781003478805-3

2018, p. 78). Despite its magnitude, the estimate later became a goal of national policy (Reinertsen and Asdal, 2019).

Where and how are such visions made? What are the mediating instruments through which new economies, such as the seaweed economy, are made? And more concretely, how can a seaweed economy, as part of the larger blue bioeconomy, be made to move offshore—far away from the wild-growing seaweed's natural habitats which are near the coast? How did the High North Atlantic become a site for cultivation?

This happens, we argue, in and through documents. In these seemingly mundane material and semiotic artefacts, mediations occur. Documents are indispensable sites for growing the blue bioeconomy, for bringing the seaweed to the economy, and for moving the seaweed economy to the High North Atlantic. In pursuing this argument, we build on earlier research on the role of documents in the making of the ocean economy (Asdal, 2015; Reinertsen and Asdal, 2019; Asdal and Cointe, 2021; Asdal and Huse, 2023). Our chapter builds upon the practice-oriented document analysis that approaches documents as sites and tools for action (Asdal, 2015; Asdal and Reinertsen, 2022). Documents not only describe and narrate things in an outside world, they actively partake in making and transforming this world. As such, the chapter is situated within the field of science and technology studies (STS) and the tradition of actor-network theory where a key element is that of attending to the performative aspects of what have been termed "literary inscriptions" and "inscription devices" (Latour, 1987; Latour and Woolgar, 1979; see also Akrich, 1992). One of the elements in practice-oriented document analysis is to attend to both the material and textual sides of the artefacts we study. Documents are tools that move things around with them, yet they may also act upon and modify the world through their word-producing capacities.

At the core of this analysis are three documents. Although different, they share important traits. They are all developed by researchers, they all address how public authorities can prepare the ground for ocean growth, and they are all institutionally affiliated with SINTEF, a Norwegian institute for applied sciences which is oriented towards "creat[ing] innovation" by conducting commissioned research for both the private and public sector (SINTEF, n.d.). Yet what interests us here is not so much the broader context, or SINTEF as such, or the interests and motivations of individual authors. Rather we focus on the documents they produce and what these do for the blue bioeconomy. In fact, being cautious of context is part and parcel of practice-oriented document analysis and the actor-network theory which inspires it. Instead of fixating on where actors and artefacts are embedded, the main focus is on how the world is being performed and enacted (Callon and Law, 1982; Felski, 2011; Asdal and Moser, 2012; Greenwood, 2024).

Obviously, our chapter is not a study of the seaweed cultivation process in its entirety. The documents we have chosen to analyze should be understood as three moments in a process—three interlinked sites—where large-scale seaweed cultivation emerges as an opportunity, and where potential challenges are presented and potential solutions are devised. Each document is a site of cultivation. Together, they enable the growth of the Norwegian seaweed industry. Moreover, they assist in moving this growth into the High North Atlantic.

On Mediations, Modifications, and Textual Devices

In line with this collected volume's interest in mediation (Alnæs, Bull, Hausken, and Sæther, this volume), we see the documents we study as involved in processes of "mediation"—a term which has also been a concern in actor-network theory, notably in the work of Latour. Latour engages with this notion in his book *Pandora's Hope* (1999). The book is perhaps best known for its conceptualization of "technical mediation", which, in short, considers how actors are modified by the relations they enter into. A man with a gun in his hand is a different actor to a man without one, just as a gun in a hand is a different actor to a gun in a drawer (Latour, 1999, p. 179). However, Latour also employs the notion of mediation to the relation between scientific discourse and the material world it refers to and works upon. There is no great abyss between discourse and material, Latour argues. Rather, the scientific object is a "circulating reference" to both the material world and scientific theory (Latour, 1999, p. 73). The concept of mediation is key to the manufacturing of this reference. The link between the scientific utterance and its referent in the material world is made through "a sequence of mediators" (Latour, 1999, p. 79).

Latour's empirical example is a group of scientists who are investigating the soil in the Amazon. The mediators are the scientists' specialized instruments such as a suitcase with cardboard boxes containing small cubes of soil and a book of color codes. Such instruments make the object knowable by the "creation of a link that did not exist before and that to some degree modifies the original two" (Latour, 1999, p. 179). The color code book, for instance, creates a link between the soil samples and the widely deployed Munsell code of colors, used to determine the exact nuance of the soil and thereby typologize it (Latour, 1999, p. 58). The soil sample is modified, turned into a number in the code and, by this modification, enabled to move from the Amazon forest into the realm of the Munsell color code through which it can be compared to other soil samples. Mediators make the scientific objects knowable, and in doing so, modify the object and enable it to move. The Amazon is transformed into samples of soil and plants, and these samples are typologized, codified, and systematized, enabling them to be turned into diagrams, arguments, and texts. Step by step, the forest is modified and thereby enabled to move into scientific discourse (Latour, 1999, p. 79). It is this version of mediation—consisting of modifications and movements—that we analyze in our chapter, and which, in line with the method of practice-oriented document analysis, we bring to bear on documents.

The mediation processes that our analysis investigates take place then, not in the Amazon, but in documents. More specifically, in documents that participate in Norwegian policy and research practices and which contribute to making the emergence of a Norwegian seaweed cultivation industry an object of both politics and research.

To show how the documents partake in these practices, we draw on two analytical concepts from the practice-oriented document analysis. First, we analyze documents as "sites", places where events unfold—that is, sites where seaweed growth is mediated, modified and moved (Asdal and Reinertsen, 2022, p. 15). Second, we analyze "textual devices" in the documents as mediators (Asdal and Reinertsen, p. 94c). Textual devices are defined as elements which "*do* things within the document text", that is to say elements which help shape the issue with which the document is concerned (Asdal and Reinertsen, p. 94). Layouts, graphs, and references are examples of textual devices in that they shape how a document is read, how narratives are crafted and visualized, and how the document creates credibility and connects itself to the world outside (Asdal and Reinertsen, p. 97). Put briefly, textual devices do work within the text—work which may influence the object the text describes.

We suggest that this analytical move offers two contributions to ocean media studies. First, that textual devices within documents can be analysed as mediators or as what Melody Jue has called "mediating apparatuses" (Jue, 2020, p. 22n79). Like the remote sensing technologies described by Jue, the textual devices we study make it possible to know the ocean—they even make it possible to describe potential futures in the ocean and to devise strategies for how to realize these potentials. Second, that by studying the mediations performed by textual devices, we get to see the meticulous work that goes into creating the bioeconomy, including the expectations and tentative ideas that may not be realized.

The empirical analysis this chapter pursues singles out three such textual devices which perform work akin to the work performed by the instruments in Latour's Amazon example: they turn the seaweed into an object of knowledge while simultaneously transforming it. The first textual device is the description of future applications of cultivated seaweed. The second textual device is an estimate of seaweed productivity. The third and final textual device is a computational model of the waters within Norway's exclusive economic zone. Just as the book of color codes in Latour's example linked the Amazon soil samples to soil science theory, our textual devices create much-needed links between seaweed, the economy, and the sea. The textual devices modify the seaweed, turning it into a cultivatable species with a significant estimated potential for growth. They also move the seaweed, in between documents, well as from its near-coastal growth zones into the High North Atlantic which becomes a site for cultivation. In this respect the movements we describe are different from those in Latour's example. While Latour describes how the forest is moved into science by turning matter into text, the movements we describe happen within and between documents—they are all textual. In our story an envisioned seaweed cultivation is moved from the realm of politics and strategy into science—and this enables the cultivation of seaweed to be moved into a scientific representation of an actual ocean, the High North Atlantic.

Document Site One: Manufacturing Growth

The first document site we will examine is the report "Value created from productive oceans in 2050" (DKNVS and NTVA, 2012).[4] This is a report commissioned by the two Norwegian scientific societies, the Royal Norwegian Society of Sciences and Letters and the Norwegian Academy of Technological Sciences – in Norwegian abbreviated as DKNVS and NTVA. The scientific societies appointed a working group with members from across the industry, politics, and the sciences. They also gave the group a mandate, which asked the group to address "scenarios" as well as the "opportunities and challenges Norway [would] encounter" in pursuing a "significant growth potential" in the value-creation of Norway's biomarine resources (DKNVS and NTVA, 2012, p. 13).

The working group was equipped with the department "SINTEF Fisheries and Aquaculture" as its secretariat, and this is also where three of the report's four authors were employed. However, the report is no ordinary research report. It is a strategy report, a hybrid document working across the genres of politics and science. According to its own description, it is one of several "strategic projects," including a governmental white paper and a marine research and development strategy (DKNVS and NTVA, 2012, p. 4). These projects all had a specific purpose, namely "to address how Norway [was] to maintain future economic growth above and beyond exploitation of its oil and gas resources" (DKNVS and NTVA, 2012, p. 4). Thus, the report explicitly addresses Norwegian politicians, urging them to "commit themselves to the development of future marine-based industries so that we can generate greater value from the resources to which we already have access" (DKNVS and NTVA, 2012, p. 12, see also Reinertsen and Asdal, 2019).

Seaweed cultivation is one of the marine-based industries the report engages with. At the time of its writing, this was indeed a future industry as there was no Norwegian seaweed cultivation industry. In fact, commercial seaweed cultivation was not even legal. Norway did have a well-established industry of seaweed wild-harvesting, producing a relatively stable yield of between 100,000 and 200,000 tonnes of wild-growing seaweed annually (Ferdouse et al., 2018. p. 77). Yet, the potential for growth was observed to lie in following the Southeast Asian example and venturing into seaweed *cultivation*. This, the report states, would enable Norway to "produce much greater volumes" in the future (DKNVS and NTVA, 2012, p. 72). Therefore, the report does not engage with Norway's well-established and relatively large seaweed harvesting industry, but exclusively with the "embryonic" industry of seaweed cultivation with a large potential for growth.

The shift from harvesting to cultivation is important as it also modifies the seaweed itself. Through cultivation, seaweed can become "a biomass containing tailored properties which in turn can be exploited for a wide range of products" (DKNVS and NTVA, 2012, p. 72). The cultivated seaweed becomes flexible, not only in terms of its biological properties, but in an economic sense too: cultivated seaweed is, in the report's wording, a "raw material with a multitude of applications" (DKNVS and NTVA, 2012, p. 72). The term "applications" is not defined in the report, but it is employed to describe possible future utilizations of seaweed. In a two-page section, the report categorizes cultivated seaweed into four such "applications": food, feed, chemicals, and energy (DKNVS and NTVA, 2012, pp. 72–74). These applications work as textual devices—they allow for a description of how the seaweed might be utilized in the future, and, since seaweed might be exploited in several ways, several categories of seaweed applications emerge. Both features are central to the report's enactment of the cultivated seaweed's potential for growth.

To take the energy application as an example. Energy, the report states, is probably "the least profitable" seaweed application. Yet it holds "major potential" (DKNVS and NTVA, 2012, p. 73). How can this be? The potential, we learn, derives from an expected growing demand for biofuels, mainly driven by the transport sector (DKNVS and NTVA, 2012, p. 73). Seaweed contains high amounts of fermentable sugars, and therefore biomass from cultivated seaweed can be turned into several different biofuels. Moreover, cultivated seaweed is likely to perform better as a biofuel than plants from agriculture. Seaweed grows faster and takes up more carbon than land-growing plants, and it is independent of soil, fresh water, pesticides, and fertilizers such as phosphorous. The last is emphasized as a "crucial factor" for making cultivated seaweed a source of biofuel, because land-based biomasses are phosphorous-dependent and "the planet's established sources of phosphorus are beginning to run out" (DKNVS and NTVA, 2012., p. 73). For the energy application to become profitable, the seaweed must be tailored. Its cultivation should be on industrial scale, cultivation strategies should further increase its sugar content, and all the biomass—not only the sugary components—should be exploited economically (DKNVS and NTVA, 2012, p. 73).

The application works in similar ways when it categorizes seaweed as food: it enables a vision for future growth. Yet, this application mediates the seaweed quite differently. At the outset it seems as if the potential is limited since Norwegians "have very little tradition … for eating seaweeds" (DKNVS and NTVA, 2012, p. 72). However, seeing the growing popularity of sushi as a sign of an ongoing change in Norwegians' eating habits, the report foresees that the "interest in seaweeds … will increase rapidly in the period leading up to 2050" (DKNVS and NTVA, 2012, p. 72). In contrast to seaweed as energy, seaweed as food is not valued as a sugary biomass. Here, the attraction stems from the variety of *different* seaweed species containing an array of nutrients such as carbohydrates, proteins, vitamins, antioxidants, and minerals, from the species' different sensory qualities, and from some species' "well-established fields of application and export opportunities" in East Asia (DKNVS and NTVA, 2012, p. 72).

The applications are interesting not only for their way of multiplying seaweed and thus cultivating its potential. They are also interesting because they allow for envisaging massive growth without first conducting a market analysis. In fact, the two-page long section about seaweed applications does not mention the market or the economy once. Moreover, the applications move the seaweed's potential utilizations to a future where interests and demands are deemed to be higher than in the report's present. In other words, the applications are mediators, which link the cultivated seaweed to the future, thus enacting large potential for growth.

So what precisely characterizes the mediation that the applications enact? What we suggest is that it is actually performing a relational process whereby both the linked entities—that is, the seaweed and the future it is to be applied in—are modified simultaneously. Seaweed as energy relies on an enhancement of the seaweed's sugar content achieved through cultivation and a transport sector fueled by biomass. Seaweed as food, by contrast, takes the seaweed to be a nutritious and tasty variety of different species, and relies on a drastic change in Norwegian eating habits. Thus, to repeat, the seaweed and the future are modified simultaneously. Mediation is a relational affair.

While the above is about establishing the growth potential qualitatively, the report also establishes the growth potential quantitatively: an annual production of 20 million tonnes cultivated seaweed with a turnover value 40 billion NOK[5] by 2050. These numbers are premised on a combination of the above established applications of seaweed. The large production volume presupposes that "[t]he exploitation of macroalgae for applications requiring high production volumes [is] maximised" (DKNVS and NTVA, 2012, p. 75). The estimated turnover value is calculated by estimating a price of 2 NOK—roughly 26 eurocents—per kilo of seaweed. This price is conditioned upon the expectation that some amount of the seaweed will be sold as biofuels, but the remainder will be sold for a higher price than biofuels (DKNVS and NTVA, 2012, p. 75). This higher price for non-biofuel applications, it is stated, "is an important prerequisite for making biomass cultivation profitable" (DKNVS and NTVA, 2012, p. 73). To sustain both a high production volume and a profitable price, then, multiple seaweed applications must be fully developed and ready for sale. The low-priced application with the high-volume demand ensures the saleability of as much as 20 million tonnes of seaweed, while the higher priced application with the smaller demand is a prerequisite for the industry's profitability.

However, there are "criteria" attached to this estimated growth potential; qualitative conditions which must be met to realize the potential (DKNVS and NTVA, 2012, p. 75). We have already mentioned some of them: the co-existence of both high-volume and high-price applications of seaweed, as well as large-scale cultivation combined with a cultivation strategy which enhances the seaweed's content of valuable sugars. Another criterion is that there will be a "universal acceptance" of the aquacultural industry occupying more space in near-coastal areas of the sea (DKNVS and NTVA, 2012, p. 74).

In earlier research, Reinertsen and Asdal (2019) have demonstrated how these "criteria for growth" effectively enhance the quantitative estimated growth potential. Posed not as challenges, but as "criteria for growth", these allow for omitting challenges from the quantitative estimate. Thereby, the estimate describes the growth potential only insofar as the criteria are met—or, to put it more bluntly, if the challenges standing in the way of growth are overcome. As we will show in the following section, the criteria for growth not only enhance the quantitative growth potential. They also direct action by pointing out which challenges to focus on in the efforts to turn the estimated growth potential into reality. As we now turn to trace the estimated growth potential into other document sites, we will show how the challenge of finding and making space for seaweed cultivation accompanies the estimated growth potential.

Document Site Two: From Potential to Prognosis

After the publication of "Value created from productive oceans in 2050", the growth estimates set by the report immediately moved on to other documents. Within a few years, they were adopted both by a governmental white paper entitled "The world's foremost seafood nation"—which was subsequently endorsed by the Norwegian parliament—and by a governmental strategy for marine research (Reinertsen and Asdal, 2019, p. 565). The second document we will delve into is a research report commissioned by the Norwegian Ministry of Fisheries and Aquaculture and funded by the state agency Innovation Norway. This document was developed within the same SINTEF department as "Value created …". Its task was to advise on which research and development areas to prioritize in order to enable a Norwegian seaweed cultivation industry to come into being. As such, it ties into the genre of commissioned research reports, from which SINTEF gains a significant amount of its income. Its title is "A new Norwegian bioeconomy based on cultivation and processing of seaweeds: Opportunities and R&D needs" (SINTEF, 2014).

Interestingly, the report's abstract and conclusion do not summarize the report and its findings. Instead, these are formulated as a "vision for a new seaweed-based bioeconomy" (SINTEF, 2014, p. 36). In this vision, cultivating seaweed "will strengthen Norway's role as the leading seafood nation as well as a leading supplier of marine, sustainable biomass" (SINTEF, 2014, p. 35). The seaweed-based economy is here enacted as something that *will* happen, and as an economy that will make Norway a "leading nation". Simultaneously, a research and development strategy is set out to direct actions towards *enabling* this future. The vision enacts a future that is simultaneously certain to happen *and* in need of carefully planned nurturing to come about.

The report makes few explicit links to the first report—"Value created from productive oceans in 2050"—analysed above. The abovementioned estimated growth potential is cited here, although presented as a "prognosis" rather than an estimate: "According to the prognosis by Olafsen et al. (2012) 4 mill tons of macroalgae biomass will be cultivated in 2030" (SINTEF, 2014, p. 17). The term "prognosis" enacts a different kind of future than the term "potential". Where the latter indicates a future which *can* happen and might be strived for, the former indicates a future which is *likely* to happen. Accordingly, it should be planned for.

The prognosticated seaweed production does indeed require planning, especially when it is placed alongside other industries in Norway's coastal areas. And this is performed by the second textual device our chapter delves into. This textual device is an estimate for seaweed "productivity" (SINTEF, 2014, p. 17). A productivity estimate may seem rather trivial, but it undertakes an important mediation: it converts production volume to an area requirement. This area requirement is "of about 250 km^2 [in 2030]. A 20 mill tons production in 2050 will require about 1 200 km^2" (SINTEF, 2014, p. 17): 1 200 km^2 is a large area, the report states, even given Norway's very long coastline, especially because the fish farming industry already takes up over 800 km^2 of Norway's near-coastal waters. In contrast to "Value created …" this report doubts that the aquacultural industries will have access "to such large areas in the future", and this might cause area conflicts between fish farming and seaweed cultivation (SINTEF, 2014, p. 17).

What we see here, we argue, is precisely the result of a mediation. While the applications in "Value created …" linked the seaweed to the future, the productivity estimate links it to the sea. With this mediation, both the future and the sea are modified simultaneously. The promising production volume has become a space problem, and the near-coastal areas begin to seem crowded.

How to manage this? The strategy is to make seaweed move. But where to? The report delineates three places that the seaweed may be moved to in the future: It may move away from the coast to offshore areas where there are no salmon farms to compete with. It may be granted access to the national "salmon fjords", where environmental protection of wild salmon prohibits the presence of farmed salmon (SINTEF, 2014, p. 18). Or it may move *into* the salmon farms, where it can benefit from the fishes' emission of nitrogen and ammonium. No decisions are made—after all, the report's purpose is not to decide, but to map out the research needed to make its vision of a Norwegian bioeconomy based on seaweed cultivation a reality. Thus instead, the possible futures are repositioned as *needs for knowledge* to be filled by basic and applied research projects. One of these needs regards "cultivation site criteria", that is, a scientific "description of the critical values and limits" that define a good site for seaweed cultivation (SINTEF, 2014, p. 17). Let us now turn to how the seaweed is made to move into the field of peer-reviewed research and how, as part of this process, the seaweed is moved offshore, into the High North Atlantic.

Document Site Three: Moving the Future into the High North Atlantic

Two years after the publication of "A New Norwegian Bioeconomy …", the Research Council of Norway launched a research programme intended to be "a central means in the Research Council's venture into blue bioeconomy" (Forskningsrådet, 2016, p. 4, authors' translation). With 150 million NOK—approximately 14 million EUR[6]—at its disposal annually, it was a large programme. Its focus was explicitly on supporting research to help Norway "meet the ambitious growth targets". The production of low-trophic marine species such as seaweed was one of the programme's prioritized themes, with a special emphasis on enabling large-scale industry and knowledge about its area requirement (Forskningsrådet, 2016, p. 31).

This is where we come to our third and last document, a peer-reviewed article entitled "The Kelp Cultivation Potential in Coastal and Offshore Regions of Norway" (Broch et al., 2019), which was an outcome of two of the research projects funded by the research programme HAVBRUK2. Here, the seaweed industry's estimated growth potential moves into the realm of science, while the cultivated seaweed is moved further offshore, into the High North Atlantic.

As recommended by "A New Norwegian Bioeconomy …", this article grapples with the problems of finding space and locating good sites for seaweed cultivation. It also addresses competition for space in the coastal zone as a problem to be managed by the Norwegian authorities. The article presents a solution to this problem—and this solution, a mathematical model, is the third textual device we will analyze. The mathematical model is described as a tool for public management, developed to aid decisions about where cultivation sites are to be located while simultaneously solving sea area conflicts. In short, it is reasoned, the mathematical model will be able to determine where in the sea the seaweed potential will be largest.

In "The Kelp Cultivation Potential …", then, the seaweed is mediated by a new mediator: a mathematical model of the sea within Norway's exclusive economic zone called SINMOD. SINMOD is, in the language of the article, a "coupled 3 dimensional hydrodynamic-biogeochemical model system", or, more colloquially, a computational model which describes the currents, temperatures, light, and nutrient as well zooplankton concentrations at different locations and over the course of a year (Broch et al., 2019, p. 2). This mediation modifies the seaweed too, whose growth is calculated as a function of exactly the factors described by SINMOD. When joined, these two models form the so-called "coupled 3D hydrodynamic-biogeochemical-kelp model system", which can compute the seaweed biomass production potential at different geographical locations. Thereby, the model can compute the optimal sites for seaweed cultivation, that is, the places where cultivation would give the highest yield.

This moves the potential seaweed growth in several ways. First, it moves it into a sea which is made to represent the "actual" sea (Broch et al., 2019, p. 10). The model's predictions are "realistic", it is reasoned, because the results of its computations correspond to the results of two cultivation trials and observations of wild-growing seaweed (Broch et al., 2019, p. 10). Therefore, the model has the capacity "to identify actual locations suitable for kelp cultivation" (Broch et al., 2019, p. 10). The term "actual locations" is important here, as it helps move the results into a new context. The model not only calculates the growth of the sugar kelp—it can also assist in moving it from the realm of the modeled to actual and assumed good locations in the ocean.

Something more happens. The seaweed is moved offshore. The model shows that offshore locations provide the best conditions for seaweed growth (Broch et al., 2019, p. 11). This is a somewhat surprising result, given that no wild seaweed grows here. Offshore, the sea is too deep and the seabed too dark for wild seaweed to grow. In the modeled ocean and for the modeled seaweed, this is not a hindrance. Since seaweed is grown on cultivation facilities just below the surface, the model only takes the top layers of the water column into account in its calculations of seaweed productivity. Thus, the model moves the seaweed into an ocean without a seabed, where the seaweed can expand beyond its usual, near-coastal growth zones and into oceanic waters. Here, the nutrients are plenty, the currents strong, and the temperatures stable—all factors which contribute positively to seaweed growth. The result of this mediation, which is also the main finding of the article, is that "the potential [yield] seemed to be higher offshore than nearshore" (Broch et al., 2019, p. 11).

The estimated growth potential that was the outcome of "Value created …" is also present—and indeed very active—in this document site. This presence is not in the sections describing materials, methods, and results. In these sections, the article follows the rigorous procedures of scientific writing, by carefully describing procedures, defining values and assumptions, citing the scientific publications where they originate, and—as mentioned above—discussing the validity of its model and the results of its calculations. The introductory and discussion sections, however, are guided by less restrictive procedures. Following quite standard conventions, they alert the reader to the article's wider relevance. These sections cite the estimated growth potential.

By vaguely describing it as a potential that "has been suggested", the circumstances of its production are left opaque (Broch et al., 2019, pp. 1, 13). Nevertheless, it acts. The potential future value and volume of seaweed cultivation in Norway positions the industry as one that can help "meet the future challenges of limited terrestrial food resources" for a growing world population, but also as one of "many sectors and interests competing for space in coastal regions". It is precisely the latter which necessitates tools for sea area management such as the model developed in the article (Broch et al., 2019, p. 2). Thus, the growth estimate becomes the premise of the model's societal relevance.

In the ensuing discussion, the quantified estimated growth potential acts as a moving force. Here, the estimated volume of seaweed biomass produced in 2050 is converted into an area requirement "in the range of 2,700–3,000 km^2"—a much larger area than the 1,200 km^2 proposed by "A New

Norwegian Bioeconomy ...” (Broch et al., 2019, p. 13). And while, the article notes, Norway’s coastal zone is large, there are already a “multitude of interests competing for space” in this area (Broch et al., 2019: p. 13). Consequently, the coastal zone stands out as an already crowded space in need of careful management—preferably using scientifically informed tools such as the model the article develops. And since the outcome of the modeling has shown a high growth potential in the areas *away* from the coast, the article concludes with strong and unequivocal advice: “Future possibilities lie in moving [seaweed cultivation] from the coastal region into the ocean”. Finally, the seaweed is sited in the High North Atlantic.

Mediations in and Through Documents

That the High North Atlantic is turned into a site for growing the nascent Norwegian seaweed cultivation industry might be analyzed as an instance of what Asdal and Huse have called the “great economization of the ocean” (Asdal and Huse, 2023). The phrase refers to an intense and ongoing process in which the ocean is brought into the economy and exploited in thorough, diverse, and increasingly intensive ways (Asdal and Huse, p. 3). The documents about the Norwegian seaweed cultivation, which we have studied in this chapter, are one example of how this economizsation can come about.

The three documents we have examined here have all described large-scale seaweed cultivation as possessing a great potential, not only for economic growth, but also for replacing the “black” carbon economy with a normatively better “good economy” (Asdal et al., 2021). As this chapter has illustrated, the Norwegian seaweed cultivation industry is envisaged as having the potential to form the basis of a “green” bioeconomy. We have shown how this potential has been enacted, *quantitatively* as an estimated growth potential, and *qualitatively* as opportunities for fueling the transport sector, thus making Norway a leading seafood nation and feeding a growing world population by cultivating an undemanding marine organism. At the same time, we have analyzed how certain prerequisites—formulated as “criteria for growth” and “research needs”—were made integral to the abovementioned potential. While the grand potentials provide good reasons for pursuing the growth of the seaweed cultivation industry, the criteria for growth provide calls for action. By making large-scale seaweed cultivation first an object for politics and second an object for research, the documents partake in manufacturing a new and green economy at sea: a blue bioeconomy. There is a methodological point to this: a contribution from the practice-oriented document analysis to ocean media studies is to consider documents as sites. In our case, documents are sites where “the great economization of the ocean” is enabled and made worth pursuing. By implication, they are actively involved in relating to and changing both the economy and the ocean.

There is another, and related, methodological point to this. Our analysis of the three document sites has emphasized a series of mediations within the documents, performed by textual devices. We have analyzed the mediations performed by three different textual devices—a multiplying application of seaweed, a procedure for the calculation of area requirement, and, last, a model for calculating not only the area requirement of the seaweed industry but also for finding its most optimal locations. These parts of the analysis may come across as both very technical and unnecessarily detailed. However, our argument is that accessing these technicalities is key to understanding not only *that* a large growth may take place but also for grasping the procedures that give rise to this large growth and the premises it is based on. In the mediation process we have described, the seaweed has been modified. In the first document, seaweed in the economy was a wild-growing marine organism that could be harvested. In the final document, seaweed is cultivatable, and its optimal growth environment is to be found in the High North Atlantic. And with these mediations, the seaweed—and with it, the blue bioeconomy—is moved into the ocean.

Notes

1 To be clear: while seaweed cultivation offshore as of yet is on an experimental stage, there are several companies cultivating seaweed in near-coastal areas, both in Norway and other places in Northern Europe and the US.
2 Among them, the NORTH SEA FARM 1, located on an offshore wind farm in the Dutch North Sea; the JIP Carbon Solutions located off the coast of Central Norway, and the ULTFARMS pilot projects in the Baltic and North Seas.
3 With the average exchange rates in 2012.
4 Here, we cite the report with reference to the two science academies who published it: the Royal Norwegian Society of Sciences and Letters and the Norwegian Academy of Technological Sciences (DKNVS and NTVA). Other works, including in Reinertsen and Asdal (2019), cite the report (SINTEF, 2012) with reference to the research institute that functioned as a secretariat for the working process or as (Olafsen et al., 2012) with reference to the report's authors.
5 Roughly 5.3 billion EUR, with the average exchange rates of 2012.
6 With the average exchange rate in 2016.

References

Akrich, M. (1992). The de-scription of technical objects. In W.E. Bijker and J. Law (Eds.), *Shaping technology/building society: Studies in sociotechnical change* (pp. 205–224). MIT Press.

Asdal, K. (2015). Enacting values from the sea: On innovation devices, value practices, and the co-modifications of markets and bodies in aquaculture. In I. Dussauge, C-F, Helgesson, and F. Lee (Eds.), *Value practices in the life sciences and medicine* (pp. 168–185). Oxford University Press. doi:10.1093/acprof:oso/9780199689583.003.0009.

Asdal, K. and Cointe, B. (2021). Experiments in co-modification: A relational take on the becoming of commodities and the making of market value. *Journal of Cultural Economy*, 14 (3), 280–292.

Asdal, K., Cointe, B., Hobæk, B.et al. (2021). "The good economy": A conceptual and empirical move for investigating how economies and versions of the good are entangled. *BioSocieties*, 18, 1–24 doi:10.1057/s41292-021-00245-5.

Asdal, K. and Huse, T. (2023). *Nature-made economy: Cod, capital, and the great transformation of the ocean*. MIT Press. https://direct.mit.edu/books/oa-monograph/5671/Nature-Made-EconomyCod-Capital-and-the-Great.

Asdal, K. and Moser, I. (2012). Experiments in context and contexting. *Science, Technology, & Human Values*, 37(4), 291–306. doi:10.1177/0162243912449749.

Asdal, K. and Reinertsen, H. (2022). *Doing document analysis: A practice-oriented method*. SAGE Publications.

Broch, O.J., Alver, M.O., Bekkby, T., Gundersen, H., Forbord, S., Handå, A., Skjermo, J., and Hancke, K. (2019). The kelp cultivation potential in coastal and offshore regions of Norway. *Frontiers in Marine Science*, 5, Article 529.

Callon, M. and Law, J. (1982). On interests and their transformation: Enrolment and counter-enrolment. *Social Studies of Science*, 12 (4), 615–625. doi:10.1177/030631282012004006.

Felski, R. (2011). Context stinks! *New Literary History*, 42 (4), 573–591. doi:10.1353/nlh.2011.0045.

Ferdouse, F., Løvstad Holdt, S., Smith, R., Murúa, P., and Yang, L. (2018). The global status of seaweed production, trade and utilization. *Globefish Research Programme*, 124, FAO.

Forskningsrådet (2016). *Programplan: Stort program for havbruksforskning – HAVBRUK2*. Norges Forskningsråd.

Greenwood, S. (2024). The problem of context revisited: Moving beyond the resources model. *Studies in History and Philosophy of Science*, 105, 126–137. doi:10.1016/j.shpsa.2024.05.012.

Gulowsen, J. (2000). *Bro mellom vitenskap og teknologi. SINTEF 1950–2000*. TAPIR Akademisk Forlag.

Jue, M. (2020). *Wild blue media: Thinking through seawater*. Duke University Press.

Latour, B. (1987). *Science in action: How to follow scientists and engineers through society*. Open University Press.

Latour B. (1999). *Pandora's hope: Essays on the reality of science studies*. Harvard University Press.

Latour, B., and Woolgar, S., (1979). *Laboratory life: The social construction of scientific facts*. Sage Publications.

Norderhaug, K.M., Skjermo, J., Kolstad, K., Broch, O.J., Ergon, Å., Handå, A., Horn, S.V., Lock, E.-J., and Øverland, M. (2020). Mot en ny havnæring for tare? Muligheter og utfordringer for dyrking av alger i Norge. *Fisken og havet*, 2020(5).

North Sea Farmers (n.d.). *NORTH SEA FARM 1*. Retrieved May 14,2024 from https://www.northseafarmers.org/about-nsf1.

Reinertsen, H. (2016). *Optics of evaluation. Making Norwegian foreign aid an evaluable object, 1980–1992* (Publication No. 581). Doctoral dissertation. Faculty of Social Sciences, University of Oslo.

Reinertsen, H., and Asdal, K. (2019). Calculating the blue economy: Producing trust in numbers with business tools and reflexive objectivity. *Journal of Cultural Economy*, 12 (6), 552–570.

The Royal Norwegian Society of Sciences and Letters (DKNVS) and the Norwegian Academy of Technological Sciences (NTVA) (2012). *Value created from productive oceans in 2050*. SINTEF Fisheries and Aquaculture.

SINTEF (2014). *A new Norwegian bioeconomy based on cultivation and processing of seaweeds: Opportunities and R&D needs*. SINTEF Fisheries and Aquaculture.

SINTEF (2014, Jan. 10). *New seaweed farm off the Norwegian coast seeks to remove CO_2 from the atmosphere.* SINTEF Retrieved May 14,2024 from https://www.sintef.no/en/latest-news/2024/new-seaweed-farm-off-the-norwegian-coast/.

SINTEF (n.d.). *About SINTEF—Applied research, technology and innovation.* Retrieved May 14,2024 from https://www.sintef.no/en/sintef-group/this-is-sintef/.

ULTFARMS (n.d.). *ULTFARMS PILOTS.* Retrieved May 14,2024 from https://ultfarms.eu/pilots/.

4 The Draupner Wave, or The Work of Physical Oceanography in the Age of Mechanical, Digital, and Aqueous Reproduction, High North Atlantic Edition

Stefan Helmreich

On January 1, 1995, a gargantuan wave hit a North Sea Norwegian natural gas pipeline-monitoring facility called the Draupner Platform, a complex consisting of two steel-latticed risers fastened to the seafloor in 70-meter-deep water. The structure, hosting a human-occupied riser (Draupner S) built in 1984 and an unoccupied satellite riser (Draupner E) built in 1994, served and serves as a central hub for monitoring the pressure and volume of flow through Norway's offshore gas pipelines (Figure 4.1).[1] According to one study of the 1995 event, "the wave hit the platform like a sledgehammer blow" (Hansteen, Jostad, and Tjelta, 2003, p. 74), and though the Draupner E, the riser that was hit, sustained some minor damage that would modify the mechanics of how the structure could absorb physical forces, the basic integrity of the armature held.

More noteworthy in the tale of the Draupner, however, became the wave itself—an unanticipated 25.6-meter-high peak in a sea of 12-meter waves. The wave was recorded from Draupner E using a downward-pointing, reflection-measuring laser aimed at registering the level of the surrounding sea (Figure 4.2). This sudden spike of water was the first recorded instance of what scientists call a "rogue wave"—a statistically unexpected wave, one at least twice the significant wave height of its surrounds (significant wave height is defined as the average of the tallest one-third of waves in a wave field, a measure tailored to the needs of small Allied amphibious landing craft in World War Two). So singular and important was the observation of this occurrence that the wave was given an individual appellation: the Draupner wave (other waves, particularly those beloved by surfers, have names—Pipeline in Hawaii, Mavericks in California—though those designations pick out recurring events, not once-in-a-lifetime happenings). The Draupner wave has become something of a totem, a touchstone, in physical oceanography. "Freak waves," long known to mariners, but considered by many scientists the stuff of myth, were now a confirmed reality. Their new designation as "rogue waves," emerging in the 1980s and 1990s, echoes older terms like *rogue elephant* or *rogue shark*, used by naturalists to refer to anomalous, wild, individual animals, ones that depart from the dominant socialities of

DOI: 10.4324/9781003478805-4

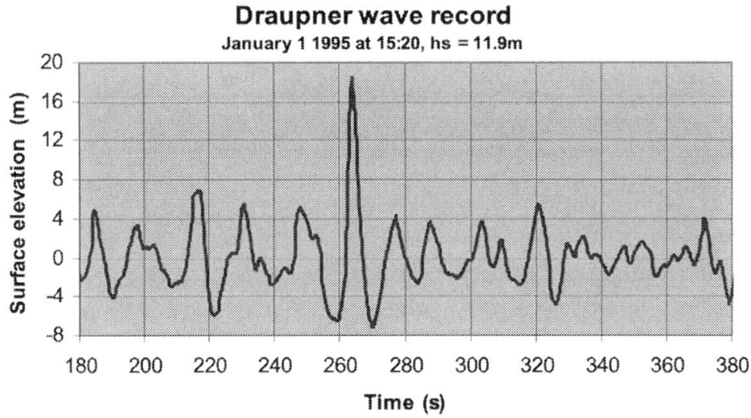

Figure 4.1 Possible Freak Wave Event Measured at the Draupner Jacket January 1, 1995
Source: Copyright: Haver, Sverre. 2004.

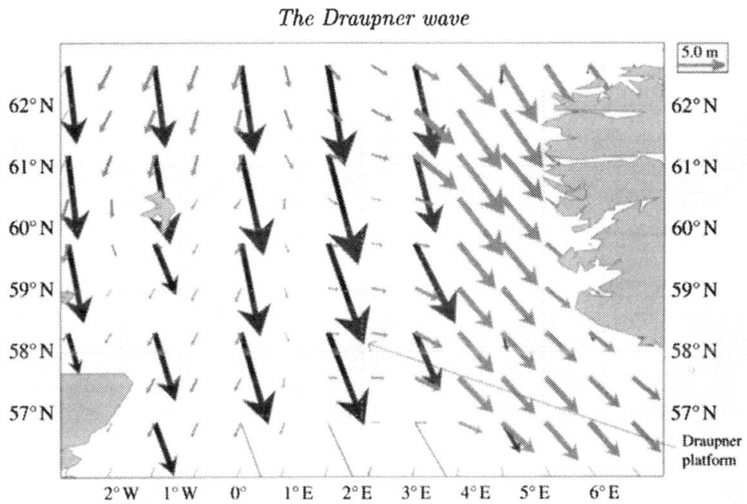

Figure 4.2 Hindcast of the wave climate in the northern North Sea at 12.00 h on January 1, 1995
Note: The north coast of Scotland and the Shetland Islands are on the left of the figure with the southwest Norwegian coast on the right. The Draupner Platform is located at 58°11'60"N 2°2'821"E as indicated. The black arrows represent the wind sea significant wave height, the grey arrows are the swell.
Source: Copyright: Thomas Adcock.

their species—less “freaks” than disobedient, asocial actors. With Draupner, a rogue had finally been captured by technical and scientific measures.

In what follows, I will examine the Draupner wave as an object—even a *work*—of physical oceanographic science. Pirating and repurposing elements of the analysis Walter Benjamin offered in his 1936 essay, "The work of art in the age of mechanical reproduction" (more recently rendered from his German into English as "The work of art in the age of its technological reproducibility"), I will examine how the Draupner wave, along with its aura, singularity, authenticity, and exemplariness, has been secured precisely through its representation and reproduction in mathematical models, computer simulations, and water wave tanks. *Reproducing* this unexpected wave offers a way both of confirming its individuality and setting it up as an example of a phenomenon that may be much more widespread. I will also argue that the Draupner's paradigmatic status, even across these many media resurrections, continues to carry something of the wave's High North Atlantic heritage.

My argument here builds upon anthropological work I have conducted among wave scientists more generally. In *A book of waves*, I draw upon ethnography I undertook among oceanographers and coastal engineers—mostly in the US and the Netherlands, though also in Australia, Japan, and Bangladesh—to examine how these researchers model, monitor, and predict waves (Helmreich, 2023). I have been interested in how these scientists represent waves as avatars of an oceanic nature at once periodic and irreversible, wild and pacific, and ephemeral and eternal. And I investigate how they do so in the field, the lab, and the computer—venues that also feature in my tale, here, of the Draupner wave. I argue that wave science, so bound up with prediction, is about *time*, about orientations toward the future—coastal, environmental, national, political, planetary. As we will see, the figure of the rogue wave serves in part as a figure of ocean futurity, since the number of rogue waves in the sea may be on the increase with climate change. But I am getting ahead of my story here. Back to Draupner.

The Draupner wave shares a name with a mythological golden arm bracelet owned by the Norse god Odin, a ringed band that generates reproductions of itself through a kind of magic. The Prose Edda reports upon the powers held by this enchanted item: "Odin laid upon the pyre the gold ring called Draupner; this quality attended it: that every ninth night there fell from it eight gold rings of equal weight." The naming of the Draupner *Platform* is easily readable as a fantasy about endlessly self-replenishing natural gas (which is, of course, actually a non-renewable, fossil fuel). It is also, of course, meant to domesticate or localize the platform as, in its way, Nordic (see Polack this volume on the naming of platforms to summon up nearby national resonances). The transfer of the name from ring to rig to wave also speaks to the generativity of the legend, the exemplar, of the Draupner wave itself. But the multiplication of representations of the wave across computer simulation, wave tanks, mathematical formalisms, science papers, and popular accounts does not deliver identical or equal waves, but rather a collection of reproductions that, I will claim, are aimed at securing the Draupner wave's authenticity. If, as Bernhard Siegert writes, the aura of the work of art "discloses itself as the origin of the work of art only in the age of mechanical reproduction" (2003, p. 30), something similar is afoot here, for the work of science in the age of reproduction across technological media—mechanical, digital, aqueous, and more. The aura for the Draupner wave, of course, does not disclose itself as the work of an authentic, originating artist (as in Benjamin's analyses), but as the singularity of a physical—and physics-described—event. The wave's aura is also revealed to be the effect of the suite of scientific tools employed to secure the phenomenon's facticity.[2]

The first question the Draupner wave posed after it was detected was about its indexicality: what did it *point* to? What had *caused* this event? Initial commentary held that the wave represented something like a 1-in-10,000-year event—though this framing did little to *explain* what had given rise to it. Was it something specific to the North Sea environment? Was the wave an example of an event that might have more general conditions of possibility? The answer would be a bit of both. For oceanographers, the Draupner wave—and all of the data collected around it—became a widely known exemplar through which to explore and theorize extreme wave phenomena, both in the North Sea and elsewhere. The tools that scientists and engineers employed included 1) the record of the event itself, 2) physical theories and mathematical formalisms describing wave action, 3) tanks of water in which to recreate the wave at miniature scale, and 4) computer simulations modeling both the weather during the Draupner event as well as the shape of specific wave forms generated by storms. The sources for apprehending the wave spanned a range of media—from the mixed media of platform steel and ambient seawater to the medium of mathematical abstraction, to the medium of wave tank water, to the medium of digital replay. The Draupner wave, in reality, reflection, and resurrection, became a multi-media kind of thing. Not Baudrillard's precession of simulacra (1983), but a Benjaminian work of science in the age of virtual reproduction that retained its aura precisely because all of its reproductions sought to certify its prior reality.

Specifying the North Sea Context

Was there something specific about the North Sea environment? Perhaps in part. In *Cold water oil: Offshore petroleum cultures*, Fiona Polack and Danine Farquharson observe that "cold water physical environments are so uniquely challenging that the International Organization of Standardization (ISO) issue specific guidelines to inform operations within them," including guidelines that require operations in these zones "to properly account for the extremities of local sea and weather conditions" (2022, p. 6). In "What conditions led to the Draupner freak wave?," Jean-Raymond Bidlot and colleagues, using high-resolution retrospective forecasts (known as hindcasts) to animate a computer simulation, argued in 2016 that "waves driven by a southward-moving polar low [off the coast of Norway] interacted with a substantial local wind-generated wave system to produce the conditions conducive to the observed large rogue wave" (and see Figure 4.3). A "polar low" is a low-pressure system "found over the ocean areas poleward of the main

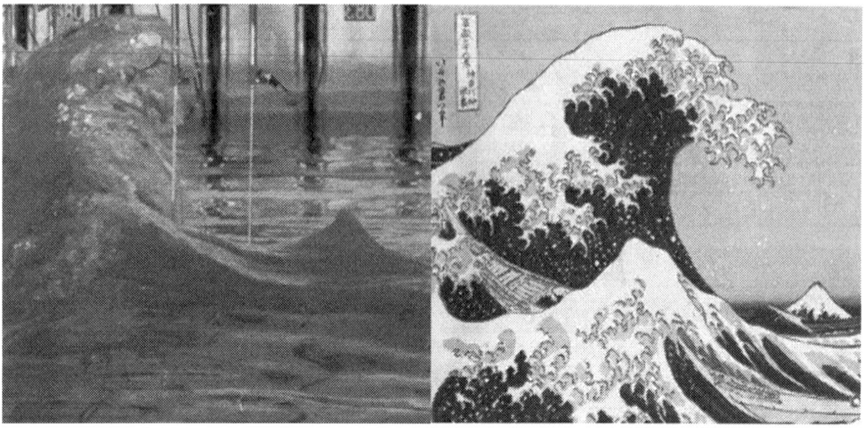

Figure 4.3 Laboratory recreation of the Draupner wave and the role of breaking in crossing seas
Source: Copyright: Mark McAllis et al., photo: Donald Noble.

polar front," and operates as a kind of "meso-scale cyclone."[3] The High North Atlantic sea is a prime spot for such phenomena, and so there is something very local about the possibility of this wave (at the same time that there are other, similar, zones such as the sea off the coast of northern Japan [see discussion below]). In the Draupner instance, the polar low, moving southward, brought cold air from the northwest. Wave expert Luigi Cavaleri and colleagues write of "the convergence in the Draupner area of two independent wave systems, … [a] large-scale one from the northwest and the one from the north associated with the polar low" (2017, p. 773).

Wave scientist Ole Johan Aarnes, at the Norwegian Meteorological Institute in Bergen, spelled out something of the North Sea specificity: "The stronger the wind, the higher the waves," Aarnes says. "If the wind is allowed to work on the sea for a long time and over a long fetch, the waves will be big." Low-pressure systems always blow in from the west to Norway, and are often amplified as they cross the North Atlantic. "This, combined with large temperature differences between different ocean currents and large areas like Greenland, makes the winds stronger and the waves extra large," Aarnes said (in Amundsen, 2018).

Air from the northwest moved in concert with southward propagating waves—though simulation results hinted that these waves were moving faster than the wind above them, suggesting that they were fed by an additional energy source—that more local wind system. The overlapping convergence of sea waves traveling in different directions, known by oceanographers as "crossing seas," seemed to be in play here (see Adcock et al., 2011). In crossing seas, waves may interfere in ways both subtractive and—important for amplifying wave heights—additive, increasing wave height. Crossing seas are by no means unique to the North Sea (though they are significant; the wave- and weather forecast Barentswatch now offers cross-wave warnings: www.barentswatch.no/en/articles/kryssende-bolger/). The Agulhas current, off the east coast of southern Africa, has long been known to amplify the waves that it meets as these swells curve around the Cape of Good Hope from the Atlantic Ocean.

The question of *where*—geospatially, on Earth—rogue waves happen has become inextricable from where they might be measured and known, turning the question from one of sheer ontology (do such waves exist?) to one entangled with epistemology (where, by whom, and with what apparatus can they be observed?). Johannes Röhrs, at Oslo's Norwegian Meteorological Institute, comments that, "There are possibly waves in the Pacific Ocean outside Japan that are as big or bigger than waves in the Norwegian Sea. *But these waves are never measured.* So the world record goes to the sea here off Norway" (quoted in Amundsen, 2018, emphasis added). This suggests that infrastructures like oil and gas platforms become crucial media for registering waves—something that wave scientist Luigi Cavaleri and colleagues (2017) have indicated should shift discussions about rogue waves' "occurrence probability" to their "encounter probability." If we wish to understand rogue waves as part of the ecology of Media Seas of the High North Atlantic, we must take into account those media—platforms like the Draupner—that are essential for revealing them in the first place. It is worth noting that the North Sea itself has long been treated as a "laboratory" for wave science; in 1973, a publication appeared entitled "Measurements of wind-wave growth and swell decay during the Joint North Sea Wave Project (JONSWAP)," which, drawing on the work of Dutch, American, English, German, and Japanese scientists, sought to generate generalizable knowledge about how and under what conditions waves grow in the open sea (see Hasselmann et al., 1973). One of the mathematicians on that project—Klaus Hasselmann—in fact was one of the first to hypothesize a mechanism by which waves (including, eventually rogues) might grow: Waves may "steal" energy from other waves, feeding those sources of power directly into their own force—a nonlinear and probabilistic process described by oceanographers using the Schrödinger equation from quantum mechanics (see Wood, 2020).

The effects of crossing seas and of nonlinear energy transfer took center stage in a physical experiment that University of Oxford engineer Mark McAllister and colleagues undertook in 2018 at the FloWave Ocean Energy Research Facility of the University of Edinburgh, a "circular wave tank surrounded by 168 wavemakers" that "allows for the creation of waves travelling in all directions" (McAllister, 2019). Unlike waves modeled in the abstract and formal realm of quantum mechanics, mechanically made waves in a tank will eventually break, not something that nonlinear Schrödinger waves do (nor something that waves investigated through analogies with nonlinear optics, which have also been forwarded to model the formation of rogue waves, do). As Benjamin writes for the work of art, so for the work of science: "technical reproduction can put the copy of the original into situations which would be out of reach for the original itself" (1936a, p. 220). Here, of course, the work of scientists was meant to fix, not remix, the Draupner's solidity as an example. McAllister and company sought to recreate in tank water, at reduced scale, the profile of the Draupner wave and they set up interfering wave trains to do so. These then converged to create a wave of exceptional (relative) height whose breaking came later than one might have expected (recall that adding wave amplitudes can push a compound wave higher than a typical wave might break). The wave also broke not with a plunge, but with a nearly vertical jetting up of water.

Wave as Icon, Emblem

McAllister and company published photos of the scale model—going on, in some venues, to argue that it also reminded them of Hokusai's woodcut print *Great Wave off Kanagawa*, which they then speculated might also have been a rogue wave.[4] The comparison is telling—and not only because, as Benjamin wrote, "With the woodcut graphic art became mechanically reproducible for the first time" (Benjamin, 1936a, p. 218), but also because with the Hokusai analogy McAllister and company make a bid to make the Draupner itself *iconic*, siphoning some of the Hokusai wave's notoriety into a bid for physical oceanographic iconicity. For communities of wave modelers, Hokusai's Wave has a certain imaginative generativity to it; reproducing it *in aqua* as McAllister claimed to do also meant to be generative of interest from popular media as well as, perhaps, future funding organizations.

The Draupner wave has not, so far, become an *icon* in quite the same way as Hokusai's Wave. Art historian Christine Guth, in *Hokusai's Great Wave: Biography of a global icon* (2015), shows how Hokusai's Wave as icon has transformed over time, from its production in late-Edo Japan to its emergence as a French, Euro-American, and then global signifier of all things Japanese; from its work in Japanese nationalism to its play in circuits of commodity capitalism; from its status as a shorthand for international high art to its popular manifestation as a lifestyle brand (think surfing); and from its early conscription into classical Western aesthetics of the sublime to its increasing association with looming global environmental disaster (and see Helmreich, 2016). The Draupner wave is different from the Kanagawa wave because the community in which its representation circulates is much more specialized than it is popular; it is a figure mostly for oceanographers.

It may serve for this community as a kind of *emblem*. Cinema theorist Kristen Whissel, writing of CGI special effects in disaster and science fiction movies, pitches the idea of an "effects emblem": "a cinematic visual effect that operates as a site of intense signification and gives stunning (and sometimes) allegorical expression to a film's key themes, anxieties, and conceptual obsessions" (2014, p. 6). The Draupner wave may perhaps be understood as a site of intense signification (a super-significant wave?) for some wave scientists. For those who understand it through the high "encounter probability" that it may have had in an oil-platformed North Sea, it may also have a particularly

Norwegian, or North-Sea-ish "encounter value" (Haraway, 2008). Will Draupner become for Norway what tsunamis are for Japan—a local wave instance that is drafted to become a universal sea phenomenon?[5] It may seem that this would be so mostly for the small professional community of wave scientists for whom its status as an exemplar matters—though the Draupner has certainly become a touchstone in a range of popular articles about rogue waves, particularly in Norway.[6] In this context, the Draupner wave may be all at once an index, icon, and emblem—a sign and symbol of that medium known as the High North Atlantic.

Coda: The Draupner Wave Meets the Angel of Ocean History

I would be remiss not to comment upon a feature that is missing in the story as I have told it here, and that is the dimension of this rogue wave as monstrous, frightening. It is important that no people were harmed during the Draupner event, that it does not live in oceanographic memory as a life-taking disaster (cf. Polack this volume, on oil platforms and calamity). The fact that the wave hit Draupner E, the unstaffed and uninhabited piece of the Draupner Platform pair, is significant—the inhuman force of the wave hit a non-human detector, attenuating associations with the embodied and terrifying. We have here in many ways a clinical instance of a rogue wave—a species of wave that, historically, was the sort of entity that would strike fear into mariners, but that here is rendered somewhat antiseptic, washed of its horrifying dimension, turned into mere measure. This is remarkable since rogue waves, according to many oceanographers, are on the increase with global warming and the intensified storms that accompany this dynamic and so in some sense might be emblems of Anthropocene fear (Helmreich, 2023). The Draupner takes the abstract measure of this fear and renders the wave reproducible for projects of rational at-sea management.

But the shadowing of the Draupner wave by fear may be increasingly inescapable, as such waves increase in frequency. We might, then, turn to another key Benjamin piece of writing, on the angel of history, to help us out in how to think about the rogue wave as the return of the repressed, the inhuman force of human damage. Benjamin writes:

> This is how one pictures the angel of history. His face is turned toward the past. Where we perceive a chain of events, he sees one single catastrophe which keeps piling wreckage and hurls it in front of his feet. The angel would like to stay, awaken the dead, and make whole what has been smashed. But a storm is blowing in from Paradise; it has got caught in his wings with such a violence that the angel can no longer close them. The storm irresistibly propels him into the future to which his back is turned, while the pile of debris before him grows skyward. This storm is what we call progress. (1940, p. 257)

I might rewrite, for the age of the rogue wave, as this:

> This is how one pictures the angel of ocean history. His face is turned toward the sea. Where we perceive the repetition of waves, he sees growing wave heights as a sign of oncoming catastrophe, promising future wreckage. The angel would like to sound the inundation alarm, and save people and nonhumans from watery catastrophe. But a storm is blowing in from the sea; it has caught the angel's feet in the sand, pinning him to the ground with such a violence that the angel can no longer move away from the ocean. The storm pummels him as he seeks to escape the steady pace of sea-level rise and the increasing probability of disastrous surges—all signs of extreme contemporary weather and climate. This storm is what we call the Ocean Anthropocene.[7]

The Draupner wave, then, represents the ocean as what Elizabeth DeLoughrey calls a kind of "daemonic agent, a figure with more than natural power that moves between realms" (2019, p. 150). The Draupner wave is a kind of devil meeting the Angel of Ocean history. No wonder it is so precisely studied, scrutinized, and modeled, all the better to tame any re-manifestations that come our way.

Notes

1 See https://www.norskpetroleum.no/en/production-and-exports/the-oil-and-gas-pipeline-system/

2 My argument here, then, is against Benjamin's claim that "The whole sphere of authenticity is outside technical—and, of course, not only technical—reproducibility" (1936a, p. 220); "The whole sphere of authenticity eludes technological—and of course not only technological—reproduction" (1936b, p. 21). Authenticity can in fact be secured by reproduction.

There is an important distinction to be named here—between a reproduction of a putatively singular, preexisting, object—and a reproduction of an object (such as a photographic print) that may never have had an original. Though Benjamin works to keep the two genres distinct—and the 1968 (1936a) and 2008 (1936b) translations provide different resources for doing so—Steigert's interpretation of the aura as in fact only a back-formation from technological reproducibility (2003) reveals something much more slippery here. One source for that slipperiness is the variety of processes that answer to *technological reproducibility*, each of which has its own affordances and relations to the thing reproduced (see Davis, 1995 for just one of many attempts to write "The Work of Art in the Age of Digital Reproduction," for example).

3 Also: an "'Icelandic low' is a large sustained atmospheric low-pressure center formed between Iceland and South Greenland. It often causes strong winter winds across the North Atlantic. In winter, the sea is significantly warmer than the continents. This difference is responsible for the location of the Icelandic low point, which dominates the wind circulation across the North Atlantic (the North Atlantic oscillation)." https://www.britannica.com/science/Icelandic-low

4 In "On Hokusai's *Great Wave off Kanagawa*: Localization, linearity and a rogue wave in Sub-Antarctic Waters," in *Notes and Records of the Royal Society*, physicist and optics theorist John M. Dudley, science writer and diver Véronique Sarano,

and mathematician Fréderic Dias postulate that Hokusai's image represents a rogue wave. Hokusai's wave has sometimes been taken for a representation of a tsunami—with *tsunami* a Japanese term adopted into international wave science over 100 years ago.

5 To avoid confusion: although Hokusai's wave was used in some Japanese and global media to represent a tsunami after the Tohoku tsunami of 2011, ocean scientists have long been adamant that Hokusai's wave not be so mistaken (Cartwright and Nakamura, 2009).

6 For a Norwegian examples, see www.forskning.no/fysikk-havforskning-naturvitenskap/verdens-hoyeste-bolger-er-i-havet-utenfor-norge/278415

www.nettavisen.no/livsstil/monsterbolger-havets-mest-lunefulle-fare/s/12-95-3423198511

www.forskning.no/fysikk-havforskning-naturvitenskap/verdens-hoyeste-bolger-er-i-havet-utenfor-norge/278415

7 Or, following, Steve Mentz (2015), the shipwreckocene, or Naufragecene.

References

Adcock, T.A.A., Taylor, P.H., Yan, S., Ma, Q.W., and Janssen, P.A.E.M. (2011). Did the Draupner wave occur in a crossing sea? *Proceedings: Mathematical, Physical and Engineering Sciences*, 467(2134), 3004–3021. http://www.jstor.org/stable/23058920.

Amundsen, B. (2018). World's highest waves form west of Norway. *sciencenorway.no*, April 16, https://sciencenorway.no/forskningno-norway-oceans/worlds-highest-waves-form-west-of-norway/1455407.

Baudrillard, J. (1981). *Simulations*. Translated from the French by Paul Foss, Paul Batton, and Philip Beitchman. *Semiotext(e)*, 1983.

Benjamin, W. (1936a). The work of art in the age of mechanical reproduction. Translated from the German by Harry Zohn. In H. Arendt (Ed.), *Illuminations* (pp. 217–251). Shocken.

Benjamin, W. (1936b). The work of art in the age of its technological reproducibility: second version. Translated from the German by Edmund Jephcott and Harry Zohn. In M.W. Jennings, B. Doherty, and T.Y. Levin (Eds), *The Work of Art in the Age of Its Technological Reproducibility, and Other Writings on Media* (pp. 19–55). Harvard University Press.

Benjamin, W. (1940). Theses in the philosophy of history. Translated from the German by Harry Zohn. In H. Arendt (Ed.), *Illuminations* (pp. 253–264). Shocken.

Bidlot, J-R, Cavaleri, L., Bertotti, L., Barbariol, F., Benetazzo, A., Janssen, P., and Wedi. N. (2016). What conditions led to the Draupner freak wave? *ECMWF Newsletter* 148: https://www.ecmwf.int/en/newsletter/148/meteorology/what-conditions-led-draupner-freak-wave.

Cartwright, Julyan H.E., and Nakamura, H. (2009). What kind of a wave is Hokusai's Great Wave off Kanagawa? *Notes and Records of the Royal Society*, 63(2), 119–135.

Cavaleri, L., Benetazzo, A., Barbariol, F., Bidlot, J.-R., and Janssen, P.A.E.M. (2017). The Draupner event. *Bulletin of the American Meteorological Society*, 98(4), 729–736.

Davis, D. (1995). The work of art in the age of digital reproduction. *Leonardo*, 28(5), 381–386.

DeLoughrey, E. (2019). *Allegories of the Anthropocene*. Duke University Press.

Dudley, J. M., Sarano, V., and Dias, F. (2013). On Hokusai's Great Wave off Kanagawa: Localization, linearity and a rogue wave in sub-Antarctic waters. *Notes and Records of the Royal Society*, 67(2), 159–164.

Guth, C.M.E. (2015). *Hokusai's Great Wave: Biography of a global icon*. University of Hawai'i Press.

Hansteen, O.E., Jostad, H.P., and Tjelta, T.I. (2003). Observed platform response to a "monster" wave. In F. Myrvoll (Ed.), *Field measurements in geomechanics: Proceedings of the Sixth International Symposium on Field Measurements in Geomechanics* (pp. 73–86). A.A. Balkema Publishers. https://www.google.com/books/edition/Field_Measurements_in_Geomechanics/V40gBCU53nMC?hl=en&gbpv=1&pg=PA73&printsec=frontcover.

Haraway, D. (2008). *When species meet*. University of Minnesota Press.

Hasselmann, K., Barnett, T.P., Bouws, E., and Carlson, H. (1973). Measurements of wind-wave growth and swell decay during the Joint North Sea Wave Project (JONSWAP). Deutsches Hydrographisches Institut.

Haver, S. (2004). A possible freak wave event measured at the Draupner Jacket, January 1, 1995. In M. Olagon and M. Prevosto (Eds.), *Rogue Waves 2004, Proceedings of a workshop in Brest, France*. Infremer.

Helmreich, S. (2016). Hokusai's Great Wave enters the Anthropocene. *Environmental Humanities*, 7(1), 203–217.

Helmreich, S. (2023). *A book of waves*. Duke University Press.

Levine, A.G. (2018). This month in physics history: January 1, 1995: Confirmation of the existence of rogue waves, *APSNews, 27(*1). https://www.aps.org/publications/apsnews/201801/history.cfm.

McAllister, M., Draycott, S., Adcock, T., Taylor, P., and Van den Bremer, T. (2019). Laboratory recreation of the Draupner wave and the role of breaking in crossing seas. *Journal of Fluid Mechanics*, 860, 767–786. doi:10.1017/jfm.2018.886.

McAllister, M. (2019). How scientists recreated a monster wave that looks like Hokusai's famous image. *The Conversation*, January 28, https://theconversation.com/how-scientists-recreated-a-monster-wave-that-looks-like-hokusais-famous-image-110304.

Mentz, S. (2015). *Shipwreck modernity: Ecologies of globalization, 1550–1719*. University of Minnesota Press.

Polack, F. and Farquharson, D. (Eds.) (2022). *Cold water oil: Offshore petroleum cultures*. Routledge.

Røsjø, B. and Hauge, K. (2011). Proof: Monster waves are real. *Norwegian Geotechnical Institute*. https://partner.sciencenorway.no/feature-forskningno-geography/proof-monster-waves-are-real/1435235.

Siegert, B. (2003). REPLICATION: There are no mass media. In H.U. Gumbrecht and M.J. Marrinan (Eds.), *Mapping Benjamin: The work of art in the digital age* (pp. 30–38). Stanford University Press.

University of Oxford Mathematical, Physical, and Life Sciences Division. (2019). Famous freak wave recreated in laboratory mirrors Hokusai's "Great Wave," University of Oxford Mathematical, *Physical, and Life Sciences Division News & Events*, January 23. https://www.mpls.ox.ac.uk/latest/news/famous-freak-wave-recreated-in-lab-mirrors-hokusai2019s-2018great-wave2019?f93f2278-1e0b-11ed-8ae4-06532f3bc3fc.

Whissel, K. (2014). *Spectacular digital effects: CGI and contemporary cinema*. Duke University Press.

Wood, C. (2020). The grand unified theory of rogue waves. *Quanta Magazine*, February 5. https://www.quantamagazine.org/the-grand-unified-theory-of-rogue-waves-20200205/.

5 Architectural Mediation of the Norwegian Sea

Nancy Couling

Introduction: Mediating Ocean Literacy

From the Vesterålen archipelago above Lofoten on the Norwegian coast, the Norwegian Sea (Norskehavet) reaches around 900 km northwest to Jan Mayen Island. It encompasses both largely unknown places and sites of intense human interaction located mostly on the Norwegian continental shelf, which is the extension of Norway's continental crust into the sea until it reaches such depths or geological characteristics as to become the oceanic crust (International Hydrographic Organization, 1953). The edge of the continental shelf in the Norwegian Sea is technically defined and varies in its distance from the coastline, but its geographic features include a steep slope down to a dramatic change in depth. With increased depth, traces of human activity reduce substantially, but still include for example, fishing vessels, trails of oil spills, radioactivity estimated to have come from the fire at Sellafield nuclear power station in the UK in 1957, multibeam data mining of the deep sea, and lines in the sea drawn to determine who has rights to exploit marine resources. These unevenly distributed activities are part of extended urban systems that demarcate, extract, transport, search, and pollute across vast distances in a reciprocal relationship to cities and areas of dense settlement. Urban agglomerations depend on relationships to ocean space in deeply rooted yet largely unseen ways.

This chapter discusses the impact of urbanization processes on the Norwegian Sea's unique biological and geophysical characteristics—its depth, currents, the Mid-Atlantic Ridge, its broad and productive continental shelf and its marine life—and asks how the spatial disciplines, in particular architecture, can communicate these transformations and include such dimensions into its creative production. This means not only to facilitate human access to such spaces through drawings and other forms of mediation, but also to describe possible new human/ocean relations in space and time. Oceans and seas are fluid territories with their own determining characteristics, and part of the realm humans have influenced and manipulated in a way comparable to territories on land; for architects and multi-scale designers, this perspective raises the question of how architecture could mediate the "High North

DOI: 10.4324/9781003478805-5

Atlantic," specifically the Norwegian Sea, contributing to ocean literacy and to a holistic understanding of this ocean as a cultural realm, including our role in its continuous formation and modification? Project work executed by master students in architecture in Bergen (BAS), Norway supervised by the author, exemplifies ways of providing intellectual, sensorial, and imaginative access to multi-species interactions and human interventions in the Norwegian Sea. It offers a critical appraisal of the current mediation of ocean space and its supporting power structures, but also an insight into a spatially complex, largely unknown world, establishing new connections, and aiming to "reenchant" not only the public, but also ourselves as designers and researchers.

"Ocean literacy" is the individual and collective understanding of the importance of the ocean to humankind, "understanding humanity's influence on the ocean, and the ocean's influence on humanity" (UNESCO IOC, n.d.). Changes in oceanic patterns, including warming temperatures and rising sea-levels, have now gained public attention, slowly shifting the common perception of the ocean as a material body of immeasurable capacity, to one that is reaching its spatial and ecological limits. While science plays a vital role in monitoring, calculating and making predictions for oceanic ecosystems, creating or re-establishing societal links to the ocean is also critical to implement behavioral and policy change in the face of the ocean's environmental problems, including sea-level rise. The "Ocean literacy for all" initiative is one of the UNESCO's responses to this need. Improved ocean literacy implies not only understanding the sea's own inherent systems and spaces (water masses, currents, depth, temperature etc), but also the contemporary levels of exploitation, pollution, species loss and infrastructure construction that have transformed the ocean. Since these aspects of ocean urbanization are frequently located far offshore, or within the water column itself, they are concealed from public view. Hence ocean literacy requires mediation.

The Urbanized Sea

An urbanized sea may seem like a contradiction in terms. Seas are large, complex systems, circulating according to planetary rhythms. Familiar (land-based) urban indicators such as population density or clearly defined settlement areas are not readily apparent. After the Second World War, maritime areas not only underwent accelerated levels of resource exploitation, but these areas also increasingly included fixed installations that demanded a new type of offshore territorialization—in particular for the exploitation of offshore oil. The urbanized sea hosts logistical, infrastructural, and legislative systems that serve vital functions for urban agglomerations on land, yet the direct links and interdependencies between them have mostly been neglected in the "city" focus of urban debates.[1, 2]

Within the tradition established by the geographer Claude Raffestin, where territory is considered a socially appropriated space, Bernard Debarbieux proposed an understanding of territory as a dynamic set of conditions needed to support a social community (Debarbieux, 2019). As elaborated by geographer Christian Schmid (Schmid, 2020, p. 96), this understanding of "the materiality of territory" in the French-, Italian-, and Spanish-speaking worlds, differs from the understanding of territory discussed by Stuart Elden, and prevalent in the English-speaking world, linked to sovereignty and state power (Elden, 2013). With specific reference to my analysis of the Barents and Baltic Seas, I have proposed a definition of maritime territory as formed through the interactions between geophysical, biological, and socio-economic layers (Couling, 2020). Such a fluid, relational perspective enables critical examination of both the sea territory as a spatial entity with specific characteristics and the largely hidden exploitive practices and movements within them.

From this perspective, the Norwegian Sea—and other seas around the world—can be understood as territories in their own right: as figure, rather than ground, as active subjects rather than passive objects. This reverses the land-based view of seas as peripheral to human interest and activity—as "voids" (Steinberg, 2001), or as hinterlands and positions the sea as the analytical starting point that looks back to land. The recognition of sea regions as territories is critical to the understanding of sea space in relation to urban dynamics. If the sea is understood and treated as a territory, urbanization processes can be analyzed at a range of scales and the notion of seascape in both its cultural and geophysical dimensions can be considered.

Parallel to cultural landscapes, cultural seascapes are produced by mutually transformative interactions between human activity and the natural environment. Specific characteristics of seascapes—the depth and salinity of the water, the seafloor composition, and their capacity to sustain marine life or resource-driven maritime activities—have steered human interaction. Thus dredging, trawling, dumping of waste, shipwrecks, laying of mines, detonations to clear munitions, drilling, and injections of carbon have contributed to a continuous molding and transformation of these very seascapes over time.

This "territorial approach" has roots in the ground-breaking spatial analysis published by ETH Studio Basel (2006) in "Switzerland—An Urban Portrait," where new types of urbanized zones were proposed throughout Switzerland's entire area (Diener et al., 2006). Through such a comprehensive, large-scale analysis, urban phenomena were revealed that had previously escaped attention—even in the "Alpine Fallow Lands," where the effects of urbanization processes were evident in very specific ways.

The transformation of the world's seas and oceans into urbanized territories is most clearly exemplified by the establishment of Exclusive Economic Zones through the United Nations Convention on the Law of the Sea in 1982. Initiated by the United Nations directly after the Second World War as the first comprehensive set of regulations for maritime areas increasingly at the center of international conflict around fishing rights, offshore oil, or the dumping of waste, the convention's main aim was ostensibly to maintain peace. However, the convention gave coastal nations exclusive rights to the use of offshore resources up to 200 nautical miles from their coastal baselines, thereby establishing vast new economic zones of national resource extraction which effectively diminished the High Seas, and therefore partly also the sea commons to around 64% of the marine space (Gugger et al., 2012).[3]

Exclusive Economic Zones (EEZs) are abstract spaces, mediated solely through pre-defined GPS coordinates which can be drawn on maps, but are geographically invisible. The definition of EEZs then unleashed planning processes to spatially allocate different activities within them, and the discipline of Marine Spatial Planning (MSP) emerged. Directly comparable to two-dimensional land-based planning practices, and using zoning tools that originating from this discipline, MSP is a management tool to avoid conflicts of interest within the marine space, but also largely to prioritize the economic performance of marine activities.[4] So far, MSP has been limited in its ability to convey the kinetic, temporal, multi-dimensional aspects of ocean space. I argue that MSP therefore contributes to ocean literacy only partially, and that multiple ways of reading the ocean remain outside of that discourse.

While local forms of ocean management have always existed in different contexts, MSP has been adopted by the European Union through the Directive 2014/89, which required each coastal nation to prepare spatial plans for their marine areas by March 31, 2021 (European Parliament and the Council of the European Union, 2014), and is promoted by the UNESCO Intergovernmental Oceanographic Commission (IOC) as a valuable ocean management practice. As a relatively new discipline, MSP has now developed its own literature and training programs and is beginning to address collaborative planning agendas across the invisible borders that define national EEZ's offshore. These comprehensive planning practices are further manifestations of the way the sea has been engulfed by processes of urbanization operating at unprecedented scales.[5]

More than a habitat for humans, the ocean is a vital habitat for other forms of life. The more-than-human perspective, which can represent these inhabitants, is therefore a way to "read the ocean" outside of traditional planning tools. This perspective has a long genealogy in different disciplines including STS, philosophy, geography, cultural studies, and anthropology, represented by authors such as Bruno Latour, Donna Harraway, Nigel Thrift, and Tim Ingold, respectively, but is also of vital interest to designers. More-than-human design requires an interdisciplinary approach, and is currently being explored in diverse ways, for example in terms of material flows and agency (Perraudin et al., 2023), and non-extractive relations with life-forms and resources. It has become increasingly important as offering ways of thinking out of the Anthropocene,[6] as it crosses the nature/culture divide and shifts the human-centric position to acknowledge a multitude of alliances across species and life-forms. Donna Haraway has called this "'becoming with' companion species in naturecultures."[7]

> More-than-human studies align with the building of a new post-carbon civilisation, because it abandons the notion of humanity living apart from

> and yet endangering the other-than-human and because historically and culturally it is such a widely shared, intuitive ontology that emerges spontaneously from a shared human capacity to prehend other entities in the world (Franklin, 2023, p. 4).

Western society has developed an overwhelmingly exploitative relationship to the ocean, firmly maintained by government collaboration with the maritime industries. The space of the sea has now become a premium resource for energy transition and for new forms of extraction and waste disposal, at the same time the ecological condition of the global ocean continues to decline.

The United Nations Decade of Ocean Science for Sustainable Development 2021–30 was initiated "to support efforts to reverse the cycle of decline in ocean health."[8] To address these complex challenges, an urgent collective, cross-disciplinary effort is required, which cannot be solved purely by science, technology, or the industrial sectors that have dominated the use and mediation of sea space in the post-World War II years.

The sea is not the land. While I argue that the sea has been urbanized, and that urbanization processes unfolding through and within it can be analyzed using urban design tools, it is also necessary to develop forms of mediation that respond to its unique features. Seas and oceans cover over 72% of the earth with a visually abstract surface that conceals the spatial inversion of a seascape deeper than land is high, home to a multitude of important and endangered species and continuously in flux. Territorial divisions and exploitative activities are largely invisible and vessel tracks trace only temporary surface markings. At a time when the ecological crisis is a global preoccupation, a range of narrative forms is needed to recount today's most critical stories. Architectural and urban design can contribute to telling these stories, as the examples in Part III aim to demonstrate.

Technical Mediation—the Norwegian Sea of Data

According to the definition provided by the International Hydrographic Organization in 1952, the Norwegian Sea cuts a vast rectangle of water from the southern tip of Spitzbergen to Iceland across Jan Mayen Island, then eastwards to a latitude of 61°North on the Norwegian west coast. The south western limit to the Norwegian Sea corresponds roughly to the geographic feature of the Greenland-Scotland ridge, and the western limit to the Mohn Ridge - part of the Mid-Atlantic ridge with depths of almost 4,000 m and inhabited by particular life-forms, about some of which still very little is known (see Meyer and Lidström, this volume). This definition of the sea space is independent from the legislative definitions that intersect it; the Exclusive Economic Zones of Norway, Iceland, and Denmark. The Norwegian Sea is open towards the North Sea and the Atlantic Ocean to the south, towards the Greenland Sea in the northwest, and the Barents Sea in the north-east, therefore its extent is not discernable from the surface, rather related to the profiles of water masses or to underwater features such as the abrupt change in depth between the Barents Sea and the Northern Atlantic across the continent-ocean transition zone.

This openness could be seen as contributing to the elusive nature of the High North Atlantic itself, which, as interrogated by this volume, is less to be understood as a geographic place than as a set of characteristics. Tim Ingold has suggested, for example, that "The North Is Everywhere":

> The North, in history is not so much a platform of the world stage upon which the drama of human history is enacted, as the very source from which history springs: It refers to the potential of the world itself, of its lands and oceans, to generate and sustain both human and non-human lives. History from, I contend, is the essence of North-ness (Ingold, 2020, p. 108).

Still, in science, the Norwegian Sea is technically mediated through oceanic data—serving as a basis for producing and distributing a scientific understanding of this complex space. Since geographic information from satellites first became publicly available from the 1970s onwards, Geographic Information Systems (GIS) have become an important resource for the spatial disciplines for both analytical and speculative purposes, creating, analyzing, managing, and mapping data. GIS is useful to trace the geophysical characteristics of the Norwegian Sea and databases have been established to collect information on species distribution and conditions such as temperature, currents, and bathymetry, gathered by environmental agencies, but also the military and the Norwegian Petroleum Directorat, as in the Mareano program.[9]

Scientists explain how, since 2000, new oceanic data-gathering systems including satellites and autonomous underwater vehicles have contributed to a "Data Tsunami," which is, however, not matched by methods of sharing and assessment which would contribute to a comprehensive understanding of the impact of human activities on the ocean (Brett et al., 2020). In 2017 only 6 percent of the global seabed had been mapped to modern standards; five years later this figure stood at 20 percent. UNESCO aims to have 80 percent of the seabed mapped by the end of the 2020s (UNESCO, n.d.). Since the Norwegian Sea has not been completely scanned or monitored—a highly expensive exercise—the chosen priority sites are key to the political ecology of ocean urbanization. In recent times, as energy companies steer efforts slowly away from petroleum to renewables, the Norwegian Petroleum Directorate are directing their attentions to the deep sea in pursuit of valuable minerals located there (see "Deep-Sea Dinner" in The Deep Sea). Maps reveal the extended arms of data mining into these previously unexplored areas, and in January 2024, the Norwegian government passed a bill enabling commercial deep-sea mining in the Norwegian EEZ, nominating a vast area of 280,000 km^2 in the Norwegian Sea as developable for mining licenses (Stallard, 2024).

The selective scanning and gathering of oceanic data produce selective forms of representation. The techno-scientific type of knowledge made available through GIS reflects the power structures active in the ocean and should not be understood as stand-alone; GIS maps, like all cartographic representations, convey only a partial "truth" (King, 1996). They should be critically scrutinized, assimilated, and synthesized with other forms of knowledge to be useful in design.

The Sea of Experience

Due to the industrialization and automation of the maritime sectors, humans increasingly lack habitual interaction with ocean space offshore. Since the Norwegian Sea contains only sporadic places of dense human occupation, such as accommodation rigs at oil-gas fields, offshore spaces appear increasingly autonomous, incomprehensible in terms of scale and similarly inaccessible in both visual/conceptual and physical terms. Security restrictions and distance from settlement areas prevent direct, unmediated contact. While Rachel Carson reminds us that the evidence of our oceanic origins runs through our very system—that the sodium, potassium, and calcium constituents of blood are similar to the proportions in seawater—she also acknowledges that human ancestors left the sea and cannot physically reenter without mechanical assistance (Carson, 2018). Specialist knowledge, skills, and technological aides are required to enter the oceanic realm on an individual basis, therefore this realm takes on an abstract character and remote status. I argue that these conditions have both perpetuated the commonly-held imagination of a pure and endless space articulated only by the horizon line and enabled the establishment of industrially-led development there, out of public sight (Couling, 2018).

As teachers, we have observed an overwhelming sentiment of inadequacy to address the urgency and scientific complexity of climate change issues in our architecture students.[10] What kinds of instruments and methods could mediate contemporary spatial conditions in the ocean, bring humans closer, and provide imaginative proposals for these spaces?

In their inspirational project "The Pacific Aquarium", comprising 12 models of otherwise publicly inaccessible sections of the Pacific Ocean, the architectural practice "Design Earth" reproduces critical human/ocean relations such as deep-sea mining, through the format of aquariums in a public exhibition. These serve as reproductions of selected ocean sites, scaled down like an architectural model and contained within the aquarium to enable human observation. They explain their intentions:

> Thus, the miniature channels our cultural enchantment with natural history and its propensity to make universal knowledge a personal experience and attempts to reclaim environmental externalities as an intimate part of the political constituency of the Earth (Ghosn and Jazairy, 2017).

Why would such mediation matter? The aim of my ongoing design-led research is to create perspectives on ocean space that can bridge the physical and psychological distance outlined above, and the gap between complex science and a common cultural understanding. To communicate the kinetic spaces and oscillating margins (Couling, 2024) of the High North Atlantic that house unique features—for example the currents that drive world climate and deep-sea hydrothermal vents—can serve to socio-culturally resituate this ocean basin, and the status of both these endangered elements within it. Elsewhere I have discussed the urgent phenomenon of "Losing Sea", whereby, while hardly noticing, we have lost the sea as a commons, as a socio-cultural space, and are in the throes of losing ecologically-balanced seas across the planet (Couling, 2023).[11] Society's "Sea Blindness" follows Daniel Pauly's principle of "shifting baselines"—mediated baselines which are adjusted regularly and through which we lose track of the scale of exploitation and decimation of marine life (Pauly, 1995). Architecture can illuminate these urgent contradictions occurring in dedicated ocean space, and can contribute to cultural ways of integrating this space into our understanding of hybrid ecologies and how to take care of them.

Throughout history, the mediation of ocean space has been explored by artists and writers, but as the complexities of climate change are added to the wicked problems of ocean pollution, acidification, or eutrophication, it has more recently become dominated by science. Multiple spatial scholars have asserted that the current ecological crises are accompanied by a crisis of the imagination (Sijmons and Hajer, 2020). In connection to ocean literacy, this implies that we not only cannot read and understand our oceans, but we also cannot enchant ourselves, as asserted by Ghosn and Jazairy, or imagine a way forward. Through the design of speculative scenarios, architecture can mediate—it can forge new links between science, unheard oceanic stories, and public consciousness.

Designing in the Sea

Architects make a valuable contribution to the production of the built and unbuilt environment first and foremost through the spatial imagination. Extending the design disciplines into the oceanic realm is also a relatively new field of investigative design and practice. I argue that it is imperative to extend the concept of territory "beyond terra" (Steinberg, Peters, and Stratford, 2018), in particular to oceans, to gain a clearer understanding of the economic activity and political power being exerted in such unbounded places undergoing continual transformation (Couling, 2023). Disciplines of architectural, landscape, urban, and territorial design can connect multi-species perspectives, a range of interconnected scales, critical environmental and social concerns, and the central contribution of design, and can imagine projective proposals.

The course at Bergen School of Architecture "Explorations in Ocean Space" aims to offer plural ways of approaching ocean space, encouraging dialogue between different types of knowledge and thereby enriching not only ocean literacy in students, but also understandings within current large-scale debates.[12] Through exhibitions and publications such as this volume, student work is exposed to an interdisciplinary audience and to the general public, thereby participating in a research trajectory that can produce a new type of knowledge or understanding. Becoming more literate in ocean space is one way of overcoming *Sea Blindness*, a way to develop a sense of public stewardship towards the sea and is a form of creative empowerment for architects and designers towards the other 72% of the earth's surface. Ultimately the research ambition of the course is to mediate the sea from within, rather than from an externalized, anthropocentric view. In relation to this publication, such mediation can potentially contribute insights into the identity of the Norwegian Sea as part of the High North Atlantic and engage in dialogue with media theory.

The 2021 BAS design exploration "Choreography for Norskehavet" was led by an interdisciplinary teaching team including an architect, an artist, and a marine biologist.[13] The study treated the entire sea under the territorial approach briefly outlined above from the perspective that the Norwegian Sea (Norskehavet) is evaluated as a being in good overall ecological condition.[14] Each of the 23 students first explored and documented interspecies dialogue with one chosen species, and subsequently analyzed the accumulated anthropogenic impacts of the sea as a whole.

This method aimed to gain an understanding of the sea's ecosystems and temporality *from* the sea and away from static planning processes towards a dynamic choreography. Interactions with biotic (living organisms) and abiotic (not derived from living organisms) protagonists such as currents, temperature, and salinity were included in the analysis. In a second step, we identified "invasive species" that had an influence on natural ecosystems, including all human-made artefacts occupying the sea space over time. Through the analysis of spatial dynamics resulting from these species interactions, distinct zones comprising geographical and oceanic properties as well as human activities, emerged in the Norwegian Sea. Technological limitations exist, for example in relation to oil extraction, aquaculture sites are clustered along the coast, whereas specific forms of life can be found only in certain depths and alongside hydrothermal vents—which in turn are attracting deep-sea mining ambitions.

The resulting project contributions discussed below are therefore embedded in discreet parts of the Norwegian Sea; the coastal zone, the deep sea, and Vestfjorden between the mainland and Lofoten. The experience of ocean space as mediated through the stories of protagonists in each of these zones is directly related to the specificity of the site; the perspective of a salmon as it negotiates the drastic changes to its environment through intensive aquaculture in the Trondheim area and cultural remediation of ocean mythology to include the obsolete monsters of the maritime industries in the coastal zone; how a fin-whale is imagined to mediate the changing oceanic environment over 500 years of industrial development in the Norwegian Sea in Vestfjorden; and in the Deep Sea, the powerful narrative of oceanic extraction meeting the unique and largely unknown deep sea ecosystems along the trajectory of deep-sea mining. A highly differentiated picture of the Norwegian Sea emerges.

These projects were each installed and presented using a range of media in an individual cell in the old Bergen jail in December 2021, or moved through the entire space, as in the case of the Last Salmon. The Deep-Sea Dinner and the Last Salmon were then reformatted and exhibited at the Lisbon Architecture Triennale (September–December 2022), winning a first prize in the university section (Lisbon architecture triennale, n.d.). The four project descriptions below provide a series of specific insights, which can be understood both individually and as a series of parallel worlds, co-existing and sometimes intersecting, in the Norwegian Sea. They mediate on the one hand a more comprehensive, more visceral understanding of the spatial composition of the Norwegian Sea, and on the other hand, issue a public invitation to reconnect to the sea commons as an integral part of highly urbanized Northwestern European land-sea interface.

The Coastal Zone

Aquaculture, shipping lanes, research stations, oil fields, military installations, zoning lines, birds, spawning fish, trawlers, kelp, cold-water corals, dreams, myths, and multiple marine species, all inhabit the Norwegian Sea's coastal zone, here defined by the depth from shore to the edge of the continental shelf. This highly productive area, propelled by the warm Gulf Stream, is also fully urbanized through a plethora of activities. The depth of the coastal zone is on average around 500 meters, then plunges down to 2000 m in a steep slope at the edge of the continental shelf. The coastal zone contains important fish spawning areas, current and future marine protected areas, and in particular the edge of the continental shelf (Eggakanten) is identified as a "particularly valuable and vulnerable area" by the Norwegian Sea's integrated management plan. Human interference throughout this zone is exemplified here from the perspectives of a salmon, and both lost and resurgent mythologies across chosen sites, aiming to illuminate tensions and potentials and promote future action.

The coastal zone of Trøndelag is the birthplace of modern Norwegian aquaculture (since 1971 on the local island, Hitra) and still an intense site of aquaculture production today. In addition, protected "salmon fjords" are located behind the industrialized coastal zone, meaning that wild salmon must swim past multiple salmon farms to reach their spawning grounds. While Norway's multi-million-dollar salmon farming industry is considered an export success, it is plagued by ecological problems and has transformed the nature of the salmon flesh itself.[15]

The Last Salmon—a space-specific audio/video walk—critically responds to the current state of salmon farming in the Norwegian Sea by conceptually placing the observer underwater and guiding them through the different stages of the wild salmon's life cycle, starting from the river, swimming through the fjord, entering to the open sea, and returning back to the river. Participants are visually and aurally guided through the exhibition space with the help of their mobile phones and headphones. In certain places, a dedicated installation provides insight into the narrative—a workstation where salmon is prepared set up in the real space, while the video presents the act of cutting and preparing salmon in the same location. At each stage, a narrator comments through the audio track, interpreting how the salmon itself perceives the river and sea and the drastic changes to its environment. "I am part of this place and this place is part of me."

This technique of overlaying reality experienced in real time and space and a fictive reality triggered by key spaces around the observer was inspired by artist Janet Cardiff's video walks, where "the fictional world of the film blends seamlessly with the reality of the architecture and body in motion."[16] From their phone screens, participants follow a series of pre-recorded events that make them enter and navigate the experience of a migrating wild salmon swimming through the same space. For example, in the video, the actual jail ceiling becomes the underside of the water surface, revealing boats and infrastructure. It takes on a strange muddy color in the phone as the salmon searches for his home spawning grounds, which have now changed dramatically (Figure 5.1).

Figure 5.1 The double reality design in the "Last Salmon" audio/video walk
Source: Airola and Scéau, BAS 2021.

The underwater acoustic world is particularly important as sound travels five times further in the sea, hence the industrial activities in the fjord contribute to the environmental degradation and the salmon's heightened stress levels. In this project, more than the technical mediation of mapping, which is also used to contextualize the project, the chosen method aims to bridge the human/more-than-human gap by immersing the public directly into the environment that the salmon itself experiences—based partly on scientific research and partly on imagination.

Moving to focus on the Norwegian Sea's oil and gas industries, the project "Ocean of Myths" explores the past, present, and future of the Norwegian Sea through a series of mythical stories, eventually envisioning a contemporary myth of the future. Illustrated through elaborate maps, a series of tarot cards interpreting the invented mythological figures and sculpted models representing new organic/mineral mutations evolving unnoticed on the seafloor, the project was inspired by medieval ocean cartography, in particular "Carta Marina,"[17] and a comparison of Norse and Indian oceanic mythology which revealed similar figures and narratives; all-present tensions balance between beneficial and destructive forces and escalating greed, resulting in destruction.[18]

Contemporary ocean cartography frequently leaves marine areas blank—a tendency that began to develop in the 17th century and the Age of Enlightenment. In contrast, this project draws invisibles into the map, giving the public access to elements and events taking place out of sight. Like the Carta Marina, the results are a combination of scientific knowledge, personal experience, and mythological narration.

Overexploitation of the ocean has persisted through contemporary oil and gas exploitation, nourishing the myth of infinite resources and industrial growth. Steel and concrete oil "monsters" with "supernatural" power now inhabit the Norwegian Sea—some also named after major figures of the Norse sagas[19], [20]—and new organic/mineral mutations are already forming on the seafloor, set to inhabit a dystopian future. Oilrigs receive the same graphic treatment and are placed alongside well-known figures of Indian mythology both in the maps and in the row of tarot cards. While restoring richness and vigor to ocean cartography, the figures within the maps also warn of powers at work in the Norwegian Sea. Today the same story of short-term capitalist profit rather than the ethics of justice and equality echoes the legends from both cultures (Figure 5.2).

Vestfjorden

This part of the Norwegian Sea is a deep fjord between Lofoten and the mainland, directly experienced by a group of students on a sailing field-trip:

> Suddenly a long, sleek, dark shape slipped above the surface and under again 5m from the boat. And another, and another. One turned and looked me in the eye. Are we dangerous (he thought)? Will we capsize

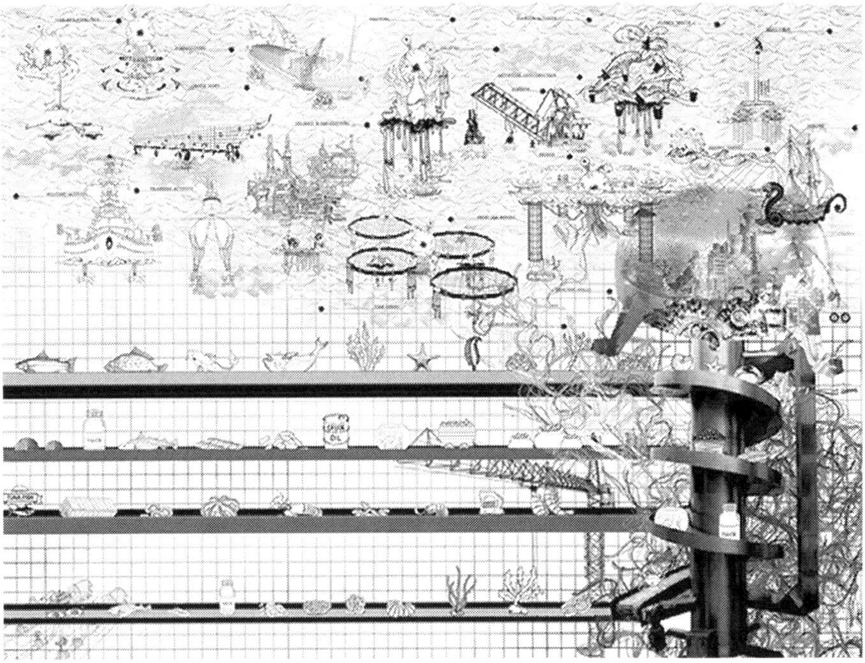

Figure 5.2 Mythological map of industrialized Trøndelag, Norwegian Sea
Source: Tiwari and Jain, BAS 2021.

> (we thought)? We are sharing this space with large, agile, intelligent creatures. They have let us in. We manoeuvre through, dependant on wind, currents and ongoing permission, exploring ocean space through physical engagement—a personal meeting.[21]

Drawing on this experience of encountering a pod of fin-whales in Vestfjorden, Magnus Gjesdal's project "*The Whale's Council*" told the story of a call to council by Hafgufa—the ancient mythological whale-god—where the fin-whale protagonist must report on the last 500 years of intensifying maritime industries and exploitation since Hafgufa himself retreated to the North Pole to escape the whaling industry. In a multi-layered 13-minute video incorporating both hand- and computer-aided drawings, the fin-whale describes the different historical epochs of technological advancement and their accompanying effects in his ocean; over time the whale observes new 18th-century vessels loaded with cannons, increasingly larger and louder vessels, and the construction of offshore oilrigs.

Talking to me? The whale asks of these new intense sounds. Shortly afterwards he releases another series of pulses … *I feel lost.* As windfarms crowd the waters alongside oilrigs, the whale sends out the message: *I can't navigate in these waters.*

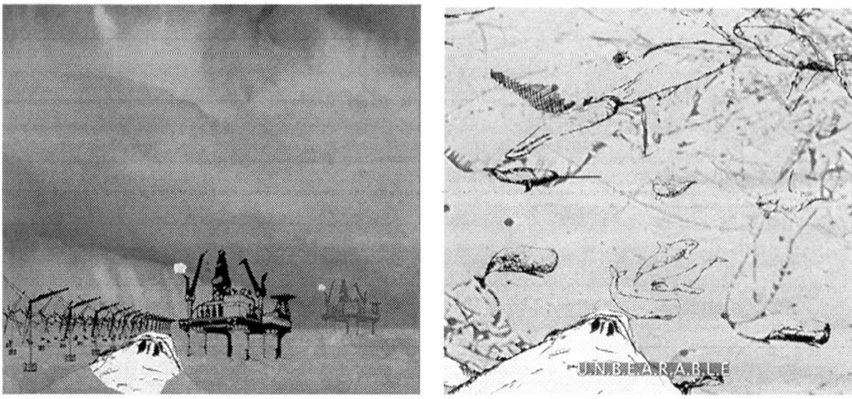

Figure 5.3 Video stills from "The Whale's Council"
Source: Gjesdal, BAS 2021.

In consultation with a marine biologist, Gjesdal's video soundtrack was conceived and composed, combining underwater recordings with a simulation of fin-whale communication in deep, low frequency staccato sounds. As the whale council came together, other whale voices are heard, including that of Hafgufa himself. The whales expressed their desperation about the unbearable levels of sonar interference in the ocean—something the video listener had also been directly exposed to (Figure 5.3).

The Deep Sea

The deep sea contains 97% of the space available for life on Earth; dark, largely unknown and full of secrets, it is described by oceanographer and underwater archaeologist Robert Ballard as "the ultimate reservoir from which life everywhere draws sustenance" (Ballard, 2000). The deep seascape of the Norwegian Sea is formed along the Mid-Atlantic Ridge, producing hydrothermal vents that are home to complex chemical interactions contributing to unique life-forms. Historically full of mythological creatures, its uncertain future is a combination of mining ambitions and slowing planetary currents; this is where the exchange takes place between the warm surface and cool deep water currents of the thermohaline circulation that drives global climate, and which is currently observed to be slowing down (Li et al., 2023).

In the project "*Deep-Sea Dinner*," the geophysical, biological, and mineral spaces of this mysterious part of the Norwegian Sea are revealed in an effort to restate its contemporary cultural and political significance. The contrasting realities of ancient, unique ocean formations as a new exploitable resource and the deep ocean as a cultural and symbolic space, are apparent. Disorientation, a dystopic atmosphere, and feelings of deep uncertainty are conveyed by the project installation, which aims to transport the public into an unknown world, where displaced consumable materials overlay future traces of human colonization.

Inspired by the irony of industry speeches at a deep-sea mining conference dinner, *Deep-Sea Dinner* is an installation that brings the deep sea close to human experience through spatial, audio and visual means. The deep sea includes the largest geological feature on earth–the Mid-Atlantic Ridge– and its unique inhabitants – still largely unresearched, and one of the last frontiers of planetary extraction (see Chapter 9 by Tirza Meyer and Susanna Lidström, this volume).

Set up in a cell of the old Bergen jail, the Deep-Sea Dinner installation evoked an abandoned future dinner party, interpreting the appropriation and consumption associated with the enthusiastic industry plans for deep-sea mining in the Norwegian Sea (Figure 5.4). The visitor hears the soundtrack still rolling- enthusiastic speeches of unimaginable profits, overlaid by deep-sea sounds. Through these sounds and magnified video views projected in the space showing strangely colored creatures, glowing minerals and smoking vents from a window, the viewer slowly becomes aware that they are deep under the sea—where we do not belong:

> We are gathered at the Deep-Sea Dinner—in a place where humans have never been- in high pressure, between active volcanos, breathing smoky water with particles of mud and microorganisms floating around us. We are eating gold and tubeworms for dinner. The melting cake of Loki's castle will be served for dessert. Questions from the future have been sent back.[22]

The unique geological and biological features of the deep sea are portrayed in a video; hydrothermal vents take minerals from magma, bringing them into the ocean as they smoke, then eventually falling as sediments onto the sea floor. Meanwhile the manganese crust is forming at a rate of few millimeters per million years. These sediments and crust are being discussed as our new oil, the next chapter of Norway's financial odyssey. In July 2021, the Norwegian oil fund surpassed 12.000 billion kroners for the first time and in January 2024 a bill was passed by the Norwegian government that will enable deep-sea mining exploration in the Norwegian Exclusive Economic Zone (see Chapter 10 by Jørgen Alnæs, this volume). Deep-sea minerals are being explored with oil technology under the pretext of meeting green energy demands, with the potential destruction of unresearched deep ocean ecosystems as a result.

Design as a Medium

While design projects have not traditionally been accepted as producing scientific knowledge, this attitude is changing as inherited research methods are being expanded in the search for new perspectives on the complex problems currently faced by science, the arts and civil society (Bathla, 2024).[23] Architect and urbanist Paola Viganò argues that the urban and territorial design

Figure 5.4 The Deep-Sea Dinner installation and supporting information
Source: Kierulf, Haukefær, Boukili, and Thoner, BAS 2022.

project has the capacity to produce valuable knowledge, in particular in the context of today's crisis with expert knowledge, and the simplification and marginalization of architectural practice (Viganò, 2016). By integrating scientific, artistic, and spatial research, these architectural projects communicate intricate relations and provide sensual access to largely unknown seascapes, making the consequences of human activities palpable.

We cannot know everything—mystery and uncertainty are an integral part of our relationship with the ocean. The Norwegian Sea is an environment operating in a state of flux where currents are shifting, thermohaline circulation is slowing down, and some fish species are moving north to cooler waters. While the planetary impact of these changes is not known, a complete portrayal of the Norwegian Sea must also mediate such uncertainty.

The methodology behind the discussed design projects—focusing on interspecies dialogue—contributed largely to the more-than-human perspectives that emerged in the results. How do the projects speak to each other? They may provide a clearer imagination of the Norwegian Sea as a pulsing space where hydrothermal vents are smoking, where whales are losing their orientation, and salmon are struggling past polluted aquaculture sites *all simultaneously*. The Norwegian Sea, and its differentiated spaces, can accommodate many singularities that, in the words of geographer Andrew Barry "do not necessarily want to be aligned" (Barry, 2018).

Then not all of the sea is readily mediated—there are gaps in knowledge and imagination. This chapter provides only shifting fragments in more and less detail, that serve to illuminate parts within a large system; the projects outlined above serve as mediating windows into places in the Norwegian Sea where specific activities are taking place between physical humans and machines, and metaphysical myths and marine creatures. These windows aim to facilitate a cultural mediation of the Norwegian Sea, rather than the solely technical, material, economic, political, or symbolic forms that have dominated the mediation of ocean space. They facilitate human participation with the sea in a fresh, imaginative way.

The projects use narrative methods to tell critical stories, describing the contextual world in which they take place, introducing protagonists and following their actions through time, and building in scientific facts, but also speculating about alternative futures. In their book *Speculative Everything*, "Design for Production" on the editors' "A" list, is challenged by "Design for Debate" on the "B" side, and "Design as Solution" by "Design as Medium" (Dunne and Raby, 2014). The projects discussed in this chapter exemplify design as a medium to understand, explore, and experience the sea in a speculative way that stimulates debate about both types of futures; those to avoid, and those to hope for.

Notes

1 For a discussion on "methodological cityism" see: Angelo and Wachsmuth (2015), "Urbanizing Urban Political Ecology."

2 See also: Couling and Hein (2020), *The Urbanisation of the Sea: From Concepts and Analysis to Design*, 9.

3 According to UNCLOS, the water column within the EEZ still functions as a sea commons, while resource extraction refers to the ocean floor, or fishing—however resource exploitation brings many spatial restrictions meaning that any "common" activities also become highly restricted and controlled.

4 The first legislative Marine Spatial Plan in Europe was completed in Germany in 2009. For more on the economic priorities of marine spatial planning, and Blue Growth, see Nancy Couling (2022), Imagining the invisible: Spatial design for the North Sea, *Planning Practice & Research*, 37(3). See also Fiona Polack, Kristin Asdal and Marie Stilling, this volume.

5 Campling and Colás call the establishment of the EEZ's "the single greatest enclosure in human history" (Campling and Colás, 2017, p. 780).

6 In February 2024, the International Union of Geological Sciences (IUGS) announced its rejection of the official proposal to formalize the Anthropocene as Earth's new geological epoch; however, it agrees that the debate has produced much valuable data on the human impact on global systems.

7 Haraway (2013, p. 32).

8 https://en.unesco.org/ocean-decade

9 https://mareano.no/en

10 In the 2022–23 academic year, BAS carried out a pedagogical experiment on non-extractive architecture across the school to address mounting psychological issues in students relating to the climate crisis.

11 This research is focused on the North Sea, where EEZs have fully consumed the space. International waters: "*areas beyond national jurisdiction*", while labelled a global common, are still subject to industries such as deep-sea mining.

12 https://explorationsinoceanspace.cargo.site/
13 Nancy Couling, architect, Vibeke Jensen, artist and Marianna Anichini, marine biologist.
14 Norwegian Ministry of the Environment (2009, p. 7).
15 At 50%, Norway is the world's largest producer of farmed salmon.
16 https://cardiffmiller.com/walks/
17 Carta Marina, by Olaus Magnus, 1539. Inscription: Marine map and Description of the Northern Lands and of their Marvels, most carefully drawn up at Venice in the year 1539 through the generous assistance of the Most Honourable Lord Hieronymo Quirino.
18 A central story in Hindu mythology is the churning of the ocean of milk, *samudra-manthan*, where gods and demons compete for the nectar of immortality, *amrita*.
19 For a fuller discussion on naming of Norwegian oilfields see Couling (2022).
20 Also see the example of Draugen (an undead creature), in Couling and Wieder (2021).
21 Project introduction as submitted to the Lisbon Architecture Triennale, 2022.
22 Project introduction as submitted to the Lisbon Architecture Triennale, 2022. Project authors: Katarina Kierulf, Bastian Haukefær, Sara Boukili, Maria Thoner.
23 For the production of knowledge through design, see also Paola Viganò (2016).

References

Angelo, H., and Wachsmuth, D. (2015). Urbanizing urban political ecology: A critique of methodological cityism. *International Journal of Urban and Regional Research*, 39(1), 16–27. doi:10.1111/1468-2427.12105.

Ballard, R. (with Hively, W.) (2000). *The eternal darkness: A personal history of deep-sea exploration*. Princeton University Press.

Barry, A. (2018, June 22). *Petrochemical zones* [Conference presentation]. Viscous Space—The offshore materiality of the North Sea between solid and liquid, TU Delft.

Bathla, N. (2024). *Researching otherwise: Pluriversal methodologies for landscape and urban studies*. gta Verlag.

Brett, A., Leape, J., Abbott, M., Sakaguchi, H., Cao, L., Chand, K.*et al.* (2020). Ocean data need a sea change to help navigate the warming world. *Nature*, 582 (7811). doi:10.1038/d41586-020-01668-z.

Campling, L., and Colás, A. (2017). Capitalism and the sea: Sovereignty, territory and appropriation in the global ocean. *Environment and Planning D: Society and Space*, 36(4), 776–794. doi:10.1177/0263775817737319.

Carson, R. (2018). *The sea around us*. Oxford University Press.

Cooking Sections. (n.d.). Salmon: A red herring. https://cooking-sections.com/CLIMAVORE-On-Tidal-Zones.

Couling, N. (2018). Formats of extended urbanisation in ocean space. In P. Horn, P. Alfaro d'Alençon, and A.C. Duarte Carduso (Eds.), *Emerging urban spaces: A planetary perspective* (pp. 149–176). Springer.

Couling, N. (2020). Ocean space and urbanisation: The case of two seas. In N. Couling and C. Hein (Eds.), *The urbanisation of the sea: From concepts and analysis to design* (pp. 189–203). nai010.

Couling, N. (2022). The offshore petroleumscape: Grids, gods and giants of the North Sea. In C. Hein (Ed.), *Oil spaces: Exploring the global petroleumscape* (pp. 109–126). Routledge.

Couling, N. (2023). Losing sea: Abstraction and loss of the commons in the North Sea. In C. Schmid and M. Topalović (Eds.), *Extended urbanisation: Tracing planetary struggles.* Birkhäuser.

Couling, N. (2024). Counter-mapping: A morphology of oscillating margins in the Norwegian Sea. In K. Peters and J. Turner (Eds.), *Ocean governance (beyond) borders: Perspectives on people and planet*. Springer Nature.

Couling, N., and Hein, C. (Eds.). (2020). *The urbanisation of the sea: From concepts and analysis to design*. nai010.

Couling, N., and Wieder, A. (Eds.). (2021). *Vestland North Sea blueprints.* Bergen Kunsthall.

Debarbieux, B. (2019). *Social imaginaries of space.* Edward Elgar.

Diener, R., Herzog, J., Meili, M., de Meuron, P., and Schmid, C. (with ETH Studio Basel, Contemporary City Institute). (2006). *Switzerland: An urban portrait.* Birkhäuser.

Dunne, A., and Raby, F. (2014). *Speculative everything: Design, fiction, and social dreaming*. MIT Press.

Elden, S. (2013). *The birth of territory.* University of Chicago Press. https://press.uchicago.edu/ucp/books/book/chicago/B/bo15506915.html.

European Parliament and the Council of the European Union. (2014, July 23). Directive 2014/89/EU establishing a framework for maritime spatial planning. European Union. https://eur-lex.europa.eu/legal-content/EN/TXT/PDF/?uri=CELEX:32014L0089&from=EN.

Franklin, A. (Ed.). (2023). *The Routledge international handbook of more-than-human studies.* Routledge. doi:10.4324/9781003262619.

Ghosn, R., and Jazairy, E.H. (2017). Leviathan in the aquarium. *Journal of Architectural Education*, 71(2), 271–279. doi:10.1080/10464883.2017.1340777.

Gugger, H., Couling, N., and Blanchard, A. (Eds.) (2012). *Barents lessons. Teaching and research in architecture.* Park Books.

Haraway, D.J. (2013). *When species meet*. University of Minnesota Press.

Ingold, T. (2020). The North is Everywhere. In T.H. Eriksen, S. Valkonen, and J. Valkonen (Eds.), *Knowing from the indigenous north: Sámi approaches to history, politics and belonging*. Routledge. https://www.routledge.com/Knowing-from-the-Indigenous-North-Sami-Approaches-to-History-Politics/Eriksen-Valkonen-Valkonen/p/book/9780367585778.

International Hydrographic Organization. (1953). *Limits of oceans and seas* (Special Publication Nr. 23). International Hydrographic Organization.

King, G. (1996). *Mapping reality: Exploration of cultural cartographies.* Macmillan Press.

Li, Q., England, M.H., Hogg, A.M., Rintoul, S.R., and Morrison, A.K. (2023). Abyssal ocean overturning slowdown and warming driven by Antarctic meltwater. *Nature*, 615(7954). doi:10.1038/s41586-023-05762-w.

Lisbon architecture triennale. (n.d.). Universities award. *Trienal de Arquitectura de Lisboa 2022*. https://2022.trienaldelisboa.com/en/evento/lisbon-triennale-millennium-bcp-universities-award/.

Norwegian Ministry of the Environment. (2009). Integrated Management of the Marine Environment of the Norwegian Sea Report No. 37 to the Storting (2008–2009) [Stortingsmelding]. https://www.regjeringen.no/en/dokumenter/report-no.-37-to-the-storting-2008-2009/id560159/.

Pauly, D. (1995). Anecdotes and the shifting baseline syndrome of fisheries. *Trends in Ecology and Evolution*, 10(10), 430.

Perraudin, L., Mareis, C., Held, M., and Winkler, C. (Eds.) (2023). *Material trajectories.* meson press.

Schmid, C. (2020). Analysing extended urbanisation: A territorial approach. In N. Couling and C. Hein (Eds.), *The urbanisation of the sea: From concepts and analysis to design* (pp. 189–203). nai010.

Sijmons, D., and Hajer, M. (2020). 2050: An energetic odyssey: Persuasion by collective immersion. In N. Couling and C. Hein (Eds.), *The urbanisation of the sea: From concepts and analysis to design.* nai010.

Stallard, E. (2024, January 9). Deep-sea mining: Norway approves controversial practice. *BBC News.* https://www.bbc.com/news/science-environment-67893808.

Steinberg, P.E. (2001). *The social construction of the ocean.* Cambridge University Press.

Steinberg, P.E., Peters, K., and Stratford, E. (Eds.). (2018). *Territory Beyond Terra.* Rowman & Littlefield International. https://www.rowmaninternational.com/buy-books/product-details/.

UNESCO. (n.d.). One Ocean Summit: UNESCO pledges to have at least 80% of the seabed mapped by 2030. UNESCO. https://www.unesco.org/en/articles/one-ocean-summit-unesco-pledges-have-least-80-seabed-mapped-2030.

UNESCO IOC. (n.d.). Ocean Literacy Portal. *Ocean Literacy Portal.* https://oceanliteracy.unesco.org/about/.

Viganò, P. (2016). *The territories of urbanism: The project as knowledge producer* (1st edition). EPFL Press.

6 Oil Media on the Grand Banks

Hibernia as Platform

Fiona Polack

> Big fear travels in the Sikorsky. Twelve-hour shifts travel with them,
> the deluge system, aqueous foam. Machinery's one note
> hammering the heart, identity compressed with intentions, drenched,
> the tired body performs delicately timed, brutal tasks no training
> adequately represents and which consume the perceivable world.
> In beds on the drilling platform in suspended disbelief,
> identified by the unlovely sea's aggression, no sleep aids,
> should a directive come. Underwater welders deeply unconscious.
> Survival suits profane in lockers. By dreams of marine flares
> and inflatables, buoyant smoke, percolating fret,
> one is weakened. Violence enters the imagination.
>
> (from Karen Solie's "Bitumen")[1]

This chapter's epigraph is taken from Canadian poet Karen Solie's disquieting reflection on the deeply compromised relationship between humans and non-humans in the petroleum age.[2] "Bitumen" opens on the Grand Banks, "east-southeast from Newfoundland toward *Hibernia*" (Solie, 2015, p. 61) and is permeated with a sense of menace. The affect-scape of Solie's poem resonates with fear-infused genres already associated with the North Atlantic offshore oil industry, including noir (Pitt Scott, 2020) and eco-horror.[3] In referencing Newfoundland's 1982 *Ocean Ranger* and the UK's 1988 *Piper Alpha* catastrophes, the poet reminds us that terror is justified in cold northern waters. She also implies that the catharsis offered by genres such as horror distracts us from confronting a now inescapably climate-changed future. As the speaker later states, in the human drive to wrench wealth from nature: "The will creates effects no will can overturn" (Solie, 2015, p. 63). Here we find ourselves in the Anthropocene, in other words.[4]

I chose to commence with these lines from "Bitumen" because they dwell so evocatively on petroleum media in the cold reaches of the North Atlantic.[5] The excerpt refers to a plethora of material technologies essential to extracting oil and gas offshore, from Sikorsky helicopters to survival suits. It also foregrounds technics, including 12-hour shifts and the abjuring of sleep medications, central to labour practices. The dissociative effects of such technologies and technics are accentuated. Significantly, while human experience is

DOI: 10.4324/9781003478805-6

foregrounded, this stanza of "Bitumen" is not focalized through an individual's perspective, and no specifically human-to-human interactions are delineated. Inversely, the nonhuman is personified and granted agency: the sea is hostile and aggressive, and machines hammer the heart. In Solie's poem, the human and nonhuman blur as the North Atlantic environment and petroleum infrastructure absorb and reflect the violence of anthropoid extractive processes.

"Bitumen" offers a corrective to the vaunting figurations of technology typical in the Canadian province of Newfoundland and Labrador's offshore petroleum industry. The poem's frank emphasis on fear is striking.[6] Atlantic Canadian oil technologies are typically promoted as infallible marvels of human ingenuity assumed to invoke awe and pride. As climate change becomes increasingly obvious and calls to halt further petroleum development grow, industry operators tenaciously adhere to this rhetoric. Proponents endeavor to deflect accusations of environmental harm by, for instance, dwelling on new methods of reducing carbon emissions on site, and claiming Atlantic Canadian oil is "green" compared to petroleum extracted elsewhere, even while they elide the downstream emissions all hydrocarbon production facilitates.[7]

The petroleum industry's technophilia is overt in the ExxonMobil-sponsored, permanent oil and gas display which opened at the Geo Centre museum in Newfoundland and Labrador's capital, St. John's, in 2022. The stridently future-facing exhibition insists the petroleum industry can devise technological innovations to master any challenge the cold water offshore poses. In a timeline of "Milestones in our oil and gas industry" (Geo Centre, 2022) the two disasters with the greatest potential to disrupt this narrative, the capsizing of the *Ocean Ranger* oil rig, which killed 84 in 1982, and the 2009 ditching of a Cougar helicopter, which resulted in 17 fatalities, are acknowledged but swiftly incorporated into the linear narrative of progress. In emphasizing how the accidents led to better safety practices, the timeline prevents the kind of "resistant mourning" Charman (2014) argues is a form of necessary recognition and protest. Moreover, accidents with no human victims, such as the 2018 spill of 250,000 litres of oil from Alberta-headquartered company Husky's Sea Rose production facility, receive no mention in the timeline at all.[8]

At the heart of the Oil and Gas Gallery is a large glass-enclosed model of Newfoundland and Labrador's first production platform, *Hibernia*. The actual *Hibernia*, one of the most gargantuan platforms in operation (Shamseer, 2023), began producing oil from the Hibernia Field in the Grand Banks' Jeanne D'Arc Basin, around 300 kilometres south-east of St. John's, in 1997. As its centrality to the Geo Centre exhibit signals, *Hibernia* is an especially compelling instance of Atlantic Canadian oil media.[9] Frequently central to the petroleum industry's self-promotional efforts, *Hibernia* has, as I detail below, always been a platform for interpellating onshore petrocultural identities as well as extracting oil. From a Media Studies perspective which approaches the extractive platform infrastructure *itself* as media, *Hibernia* is also salutary. Indeed, in what follows, I revisit Media theorist John Durham Peters' claim that the ship is "the arch-medium that reveals the ontological indiscernibility of medium and world" (Peters, 2015, p. 102). Offshore oil platforms are, I argue, even stronger contenders for the title, especially in a time of climate crisis.

The word "platform" derives from the 16th-century French *plateforme*, which, the Oxford Dictionary of English tells us, translates as "ground plan," or, literally, "flat shape" ("Platform," 2010). "Platform" has come to be deployed in multiple ways in English, most commonly as a noun referring to, first, "a raised level surface on which people and things can stand" ("Platform," 2010), for purposes as varied as obtaining a better view or performing for an audience, or, second, "the declared policy of a political party or group" ("Platform," 2010). More recently, in a Media Studies context, "platform" names a "programmable digital architecture designed to organize interactions between users" (Van Dijck, Poell, and De Waal, 2018, p. 4). Threaded through all of these definitions are notions of platforms as spaces of facilitation and connection. Platforms provide the "ground plan" upon or through which the material (oil, Airbnb accommodations, car rides), and the immaterial (affects, perspectives, ideologies) can be conveyed. Platforms (like other forms of infrastructure) are not, however, neutral. Of online platforms, Van Dijck, Poell, and De Waal comment: "platforms do not simply connect social and economic actors but fundamentally steer how they connect" (Van Dijck, Poell, and De Waal, 2018, p. 24). This claim can be broadened to other kinds of platforms—including offshore extractive ones—and to encompass environmental, as well as social and economic, concerns. On *Hibernia*, for instance, the architecture of the platform means that most workers do not interact directly with the ocean, potentially increasing their alienation from marine ecologies. The notion digital "platforms do not operate in isolation" (Poell, 2020, p. 651) also has wider application. *Hibernia* is embedded in various more-than-human assemblages, ranging from local and global communications networks, to the share market, to seabird migration patterns.[10]

Introducing a special oil-focused issue of *Media Tropes*, Jordan Kinder and Lucie Stepanik argue most scholars interested in the socio-cultural dimensions of petroleum focus on "the aesthetic or representational relationship between oil and media" (Kinder and Stepanik, 2020, p. ii) rather than "critically approaching oil and its infrastructures as media" (Kinder and Stepanik, 2020, p. iii). Centring the concept of the platform (one Kinder (2023) himself engages elsewhere) I endeavour to do both in this chapter.[11] I initially consider how *Hibernia* has been deployed as a platform for building affective attachments to oil in onshore communities. After first sketching the long history of extractive colonialism in the waters in which *Hibernia* is situated, I examine the publicity campaigns leading up to the platform's installation offshore in 1997, as well as subsequent celebrations of its billionth barrel of oil. In the latter part of the chapter, I broaden my lens to consider the ontological implications of extractive offshore platforms generally. Inspired by, but also challenging, John Durham Peters' observations about ships, I argue that petroleum media such as *Hibernia* are uniquely emblematic of more-than-human relationships in the Anthropocene.

Locating *Hibernia*

Exploitation of the Grand Banks, a series of shallow-water plateaus off Newfoundland's coast, dates back to at least the fifteenth century, when Europeans became aware of the fertile fishing grounds produced by the blending of the Gulf Stream and Labrador Current. The discovery prompted the development of a summer-based migratory transatlantic fishing industry and, eventually, British and French colonization of nearby *Ktaqmquk*/Newfoundland/*Terre-Neuve*.[12] As American historian Charles Travis and his co-authors argue, the creation of the "oceanic plantation" of the Grand Banks was "the result of a confluence between cartography, commerce, and cultural agency [and] contributed to capitalising modern European and North American societies" (Travis et al., 2020, p. 2). This offshore region became key to global European imperialism, generating wealth to fund colonial expansion, acting as a training ground for seamen, and providing an important protein source for Europeans and their African slaves. Over the past five centuries, exploitation of the Grand Banks has constituted "a bellwether for the massive resource extractions of modernity" (Travis et al., 2020, p. 3). In the late twentieth century, damaging new fishing technologies brought cod stocks to the point of collapse, and in 1992 the Canadian government imposed a moratorium on the northern cod fishery. Cod was spared extinction although, 30 years later, remains endangered (Smellie, 2021).

Coterminous with the decline of cod, and then spurred by the fishery's collapse, Newfoundland and Labrador turned to oil. As Canadian political scientist Angela Carter details, "Newfoundland's history is marked by a dual dependence on natural resource extraction and extreme environmental exploitation in pursuit of socio-economic sustainability" (Carter, 2020, p. 96). Offshore oil has become the latest chapter in this history. American petroleum company Chevron struck large reserves on the Grand Banks in 1979. However, disputes between the provincial and federal governments delayed the extraction of first oil from the Hibernia Field until 1997. Since then, five other major petroleum projects—Terra Nova, Hebron, White Rose, White Rose Expansion, and Hibernia South Extension—have been implemented on the Grand Banks (Government of NL, 2023). Oil companies and the provincial and federal governments continue to pursue expansion. Despite intense opposition from environmental activists, majority-state owned Norwegian company Equinor gained Canadian government approval for its proposed Bay du Nord development in 2022. Located 500 kilometres from the coast in 1200 metres of water, Bay du Nord is likely to extend industry into deep water by the close of the 2020s (Equinor, n.d.).

This drive to expand persists despite mounting evidence that the offshore oil industry is damaging the local marine environment. While it also, of course, contributes to carbon emissions and impacts globally, the consequences of oil extraction are increasingly visible on the Grand Banks themselves. Fisheries and Oceans Canada reports escalating climate-change damage: sea-ice coverage is decreasing, ocean currents are shifting, ocean acidification is growing exponentially, sea levels are rising on adjacent coasts, and virtually all marine species are under stress (Government of Canada, 2018).

There are other petroleum extraction-associated ecological challenges, too. Ocean plastics are pervasive across the North Atlantic (Ostle, 2019) and, on the Grand Banks alone, close to one million litres of synthetic based drilling fluid and other hydrocarbons have been released since oil production began (C-NLOPB, 2023).[13] These discharges have generally occurred in small volumes, however in 2018 a quarter of a million litres of oil, gas, and water leaked from the *Sea Rose*. As Dwayne Avery (2020) details, Husky used a plethora of effective media strategies to deflect public concern. However, a growing body of scientific research outlines the threats posed by spilled hydrocarbons to the health of North Atlantic fish.[14] Research also suggests that sustained anthropogenic noise associated with seismic testing, as well as gas flaring and light pollution, are having detrimental effects on marine life and the millions of seabirds that traverse the region (see, for example, Burke, Montevecchi, and Wiese, 2012; Fraser and Racine, 2016; Fraser and Carter, 2018; Bird, 2020).

Platform for Publicity

Many of these environmental challenges have become especially pressing in recent years. However, serious ecological concerns have been present since at least the 1980s, when inaugural environmental studies associated with *Hibernia* were conducted (Mobil Oil, 1985). Worries about the potentially negative social ramifications of reorienting Newfoundland and Labrador's economy around petroleum also date back over 40 years. Large-scale studies by Memorial University researchers on oil-producing Scotland and Norway were commissioned in the 1980s to help address the question of *Offshore Oil: Are We Ready?*, as a locally-produced documentary put it (MacLeod, 1981).[15]

The industry-led publicity campaigns centred on *Hibernia* in the 1980s and 1990s were thus oriented around persuading local publics that hydrocarbon extraction and environmental protection were compatible, and provoking affective attachments to oil. These tasks were made difficult by painfully fresh local memories of the *Ocean Ranger* tragedy. The Royal Commission into the rig's capsizing found flaws in the semi-submersible's design, grievous problems with the government's regulatory structure and reckless disregard in the petroleum industry for worker safety (Royal Commission, 1984).

Corporate shareholders ExxonMobil Canada, Chevron Canada Resources, Suncor, Canada Hibernia Holding Corporation, Murphy Oil, and Equinor

Canada currently comprise the Hibernia Management and Development Company (HMDC). It was, however, ExonnMobil's predecessor, Mobil Oil, that took the biggest role in promoting initial development. In its 1985 Environmental Impact Statement (EIS), the company noted that its "major communications program" encompassed "public attitude surveys, public and business community consultations, supplier and manufacturing capability surveys, school programs and distribution of audiovisual materials" (Mobil Oil, 1985, p. 7). Mobil also sponsored the creation of newsletters, books, videos, and media advertising campaigns, as well as operated information centres in St. John's, and at the *Hibernia* construction site at Bull Arm.

The climate crisis was not at the forefront of most people's minds in the 1980s, and the *Deepwater Horizon* disaster was yet to occur. Mobil thus had more leeway to promote the notion that oil and environmental protection were compatible. At the Bull Arm visitor centre, children were given games accentuating the oil industry's supposed ecological benignity. In one, they could colour an impression of *Hibernia* surrounded by diverse, disproportionately large, sea creatures.[16] Virtually all promotional media associated with *Hibernia* emphasize the platform's monumentality. However, the colouring sheet downplays it. The image is reminiscent of the phenomenon Dolly Jørgensen (2014) identifies of "naturalizing" offshore oil rigs by, for instance, placing scale models of the infrastructure in aquariums. However, in the physical act of colouring the platform and marine animals, children were not gazing at a museum exhibit; a more intimate, corporeally instantiated, connection was being made with oil. Moreover, the pleasure derived from play was associated with that link.

To further diffuse concerns about the oil industry, *Hibernia*'s proponents invoked Newfoundlanders' strong sense of "national" identity.[17] The very name "Hibernia" evokes local histories of seafaring, Newfoundland's globally significant role in facilitating nineteenth-century telecommunication technologies, and the island's longstanding Irish connections. As Maynard (1997) notes, the British steamship *Hibernia* assisted the *Great Eastern* in laying the first telegraph cables connecting Europe and North America in the 1870s. And, as the Latin term for Ireland, "Hibernia" also acknowledges the country from which many of Newfoundland's early white settlers came (Context Energy Examined, 2017).

Hibernia's corporate owners also emphasised Newfoundland workers' roles in constructing the platform, and accentuated close connections between onshore and offshore worlds. These approaches are pronounced in *Hibernia: Promise of Rock and Sea* (1997), a large-format, photo-illustrated, "quality keepsake chronicling the inception and development of a great enterprise." *Promise of Rock and Sea* is prefaced with (and was presumably funded by) advertisements from the companies involved in building *Hibernia*. For instance, atop a two-page photographic spread juxtaposing the platform with Bull Arm's snow-covered hills, Chevron's text states: "Credited with making the Hibernia discovery in 1979, Chevron Canada Resources offers its sincerest congratulations, thanks and gratitude to the people of Newfoundland." The publication proper combines hagiographic, aesthetically striking, images of workers at Bull Arm with key short quotes and explanatory text. The promotional film, *Hibernia: A Pioneer's Journey to the Grand Banks*, produced by Ad Vantage Productions and G.M. Media Productions in 1998, adopts similar strategies. Its narrative of technological triumph is framed by the testimony of an "everyman" Newfoundlander. This protagonist is farewelled at St. John's airport by his family as he boards a helicopter to *Hibernia* at the beginning of the film, and welcomed by them on his return at its close. As we view the opening frames, the protagonist tells us that in travelling to *Hibernia* he follows the precedent set by his fisher forefathers, who also worked at sea.

The oil industry's attempts to yoke Newfoundland's supposed past (fish) and future (oil), while sparking patriotic attachment to the latter were, however, best exemplified in the series of full-page ads the HMDC ran in the local *Evening Telegram* in 1996. Each advertisement was illustrated with maps of Newfoundland and Labrador, photographs of hard-hat clad construction workers, and the tag line: "Every Newfoundlander can take pride in *Hibernia*—it's something we should all rant and roar about." The use of the pronoun "we" merges Newfoundlanders and the oil industry. The encouragement to "rant and roar" alludes to well-known Newfoundland folk song "The Ryans and the Pittmans," composed by H.W. LeMessurier in the late nineteenth century (Murphy, 1912, p. 6). Newfoundland folk songs about seafaring were referenced on other occasions, too. The November 13, 1996 advertisement, for instance, quotes lyrics from "I's the B'y." The words of the song are juxtaposed with a list of statistics promoting the Hibernia project's supposed direct and indirect socio-economic benefits to Newfoundland.

The HMDC's appropriation of Newfoundland's folklore and its seafaring past is reminiscent of a phenomenon the late American visual artist and critic Allan Sekula (2002) documents in reflecting on Bill Gates's purchase of Winslow Homer's painting "Lost on the Grand Banks." Sekula associates Gates's 1998 act with the "absorption of once radical devices into the toolkit of corporate liberalism" (Sekula, 2002, p. 29), and stresses that the purchase made the artwork private property. While the commodification of folk songs such as the "Ryans and the Pittmans" and "I's the B'y" began (in, for instance, touristic contexts) before the HMDC deployed them, the process takes a different turn in the *Hibernia* advertisements. Here, a consortium of private companies positions itself as at one with a people, and implies it is less motivated by pursuing profit than promoting public good. The HMDC's advertisements also suggest that the very differently scaled activities and risks of offshore oil extraction are interchangeable with pre-industrial fishing practices, while paradoxically touting *Hibernia*'s technological sophistication. The campaign implies the offshore petroleum industry is fulfilling Newfoundland's aspirations for prosperity and modernity while respecting its "authentic" culture.

Notwithstanding its attempt to invoke Newfoundland nationalism specifically, this approach resonates strongly with the ways technological nationalism is routinely deployed to overcome political opposition to oil development in Canada more broadly. Media Studies scholar Darin Barney observes that resource commodity-related infrastructure is frequently "idealized not only as instrumental to the Canadian economy but, moreover, as materializing the Canadian nation" (Barney, 2017, p. 79). Barney's main focus is the construction of pipelines on the Canadian mainland. However, technological nationalism is also unquestionably in operation in the figuring of *Hibernia* as essential to the furthering of the "imagined community" (Anderson, 1983) of the Newfoundland "nation."

The HMDC's efforts to use *Hibernia* as a promotional platform could, on occasion, miscarry. In 1999, two years after first oil, Canadian journalist Daniel Wood published an extended essay in Air Canada's *En Route* magazine based on his first-hand impressions of *Hibernia*. Access to oil installations on the Grand Banks is tightly controlled and Wood's visit to the platform could only have occurred with operator approval. However, the article commences with the arresting lines:

> Up here, I'm as near to hellfire as I ever plan to be. Any closer to the 20-metre flame and I'd join some of the storm petrels that circled the huge blowtorch last night. A few, flying too close, ignited in the 1000 degree Celsius heat and fell like Icarus into the ocean far below. Between here and there, illuminated in the venting gas flare's incandescent glow, is an island called *Hibernia* (Wood, 1999, p. 48).

Wood's essay negotiates a delicate balance between implied criticism of offshore oil and the recounting of human-interest stories about *Hibernia*'s workers. It is invaluable for its rare first-person evocation of *Hibernia* as a platform for often violent conjunctions between human and nonhuman. Wood comments that "to call *Hibernia* a *strange* place would not do justice to the adjective" (Wood, 1999, p. 54). From his article emerges a portrait of an intensely paradoxical space. The platform's inhabitants are at the mercy of their environment, prone to the impact of gale-force winds, precipitation, waves and fog, and violently destructive of it; drilling is both a primitive undertaking—"At the end of the day…you're loaded with mud and drilling fluids"—and reliant on technology reminiscent of "a 21st-century space station"; the scale of *Hibernia* is overwhelming, yet being upon it produces debilitating claustrophobia; what lies beneath the seafloor is a matter of intense focus, but the sea itself and the life it contains is less so.

Nearly 25 years later, Wood's conception of the troubling entanglements of human and nonhuman at *Hibernia* remains unique in figurations of the platform. Indeed, more recent promotional mediations of *Hibernia* remain firmly focused on evoking Newfoundland nationalism and foregrounding the local onshore community's imbrication within a supposedly benign modern petroleum industry. A short promotional video made in 2016 to celebrate *Hibernia*'s billionth barrel of oil (HMDC, 2016) is directly reminiscent of *Hibernia: A Pioneer's Journey to the Grand Banks.* Images of reunited families are, for instance, again foregrounded. The unnamed director also must, however, address the platform's age. ExxonMobil's *Hebron* platform, installed on the Grand Banks in 2017, has, to an extent, usurped *Hibernia*'s supposed position at the cutting edge of technological sophistication. *Hibernia* is thus now also figured as the "quiet workhorse" (HMDC, 2016) of the offshore, even as the video also features offshore workers contemplating banks of computer screens. Such images remind us that, like all offshore infrastructure on the Grand Banks, *Hibernia* does not only produce oil but also gathers masses of data, participating in what Nick Srnicek calls "platform capitalism." As Snricek argues, "just like oil, data are a material to be extracted, refined, and used in a variety of ways" (Snricek, 2016, p. 40).[18] Elsewhere, and sometimes concurrently, *Hibernia* is repeatedly referred to not just as a "quiet workhorse" but also as "the jewel in the crown" of Newfoundland and Labrador's oil industry (see, for instance, Roberts, 2016). These contradictory evocations of humble, domestic animal labour alongside grandiosity and value beyond price inadvertently call to mind again Wood's description of *Hibernia* as a fundamentally "*strange*" place. I want now to approach this question of *Hibernia*'s strangeness from another perspective, by considering the platform itself as media.

Staging/Estranging Extractivism

Hibernia is a platform for human life and industry offshore. Its topsides were, to use the cyborg-evocative industry term, "mated" with a 450,000-ton concrete gravity base structure (GBS) at Bull Arm before the structure was towed to the Grand Banks and ballasted to rest on the seabed ("Hibernia Oil and Gas Field Project," 2021). As a place of human habitation at sea, *Hibernia* is both like and unlike the other most common maker of human "ground" on the Grand Banks: ships.[19] The nuances are important in discerning their mediational roles.

In *The Marvelous Clouds: Toward a Philosophy of Elemental Media*, John Durham Peters conducts an ocean-based thought experiment geared towards discovering "what role technology plays in our worlds, and how we might think about media, body, and being" (Peters, 2015, p. 56). His premise is that in oceanic contexts, humans (in marked contrast to cetaceans) are radically displaced and reliant on technology. Durham Peters reads ships as metaphors for our dependence on "craft." This metaphorical approach is important: Durham Peters is not much interested in nuances between different kinds of seacraft—although he does distinguish between boats, which "hug the shore" and consequently are "not outfitted as miniature worlds" and ships—or in the many ways in which ship technologies have evolved over millennia (Peters,

2015, p. 105). His larger point is that "On a ship, existence and technology are one. Your being depends radically on the craft" (Peters, 2015, p. 102). From this metaphorical perspective, we could, perhaps, include offshore rigs and platforms in Durham Peters' broadly conceived "ship" categorization. They too make a stage for human life at sea. What happens, though, if we extend Durham Peters' thought experiment in a different way and highlight category distinctions between ships and offshore oil platforms?[20] Indeed, I wish to suggest that the offshore oil platform makes an even more compelling "arch medium" than the ship in the Anthropocene, albeit one of much more recent provenance, that is far less culturally visible and storied.[21]

As is the case for sailors, the safety of those employed on offshore oil platforms is bound up in the fate of the "ground" on which they stand at sea; if something catastrophic happens to the infrastructure they, too, are in mortal peril. However, platforms offer a different order of ground from ships. Affixed to the same seabed for decades, platforms only very rarely negotiate coastal spaces, and never under their own power.[22] Offshore, they much more closely resemble *actual* ground, mitigating their inhabitants' direct negotiations with the ocean. Fixed in place, extractive platforms can disappear from onshore view. Their relative invisibility, and disinterest in their watery surroundings, make offshore platforms particularly useful analogues for the ways contemporary media infrastructures seemingly, and unobtrusively, detach themselves from place, even as they have profound environmental impacts.

It is the offshore platform's extractive traffic in hydrocarbons, however, that is most metaphorically generative. Oil workers' vulnerability is amplified by the fact they remove volatile substances from deep beneath the seafloor; they deal with more facets of "nature" than just the ocean. The nautical (albeit not navigational) and extractive-related technologies necessary to safeguard human life on an oil platform are thus even more elaborate and multitudinous than those required on a ship. In and of themselves customized examples of complex engineering,[23] platforms such as *Hibernia* also house extensive agglomerations of technologies—from drilling fluids, to personal gas monitors, to "hydrocarbon jet fire" resistant seawater lifts (Wilson and Fazel, 2022). These technologies are associated with an array of technics, too: every "genre" of technology must be operated and maintained by appropriately trained crew (see, for instance, the *Atlantic Canada Offshore Petroleum Standard Practice for the Training and Qualifications of Offshore Personnel*, 2020). Furthermore, the profound risks associated with petroleum extraction at sea, and confirmed by a series of catastrophic accidents, has also generated what anthropologist Hannah Appel identifies as "heavily choreographed exercise[s] in the performance of safety" (Appel, 2019, p. 263). Writing ethnographically about Equatorial Guinea's petroleum industry, Appel observes that offshore labour is saturated "with practices, performances, media, and bureaucracies of risk avoidance and safety–many of them written on and enacted through the body in dress and modes of walking" (Appel, 2019, p. 263). Her observations are even more relevant to platforms located in the tempestuous, seasonally ice-strewn, north-west Atlantic. *Hibernia* is ground upon which to stand, home to multiple, and complex technologies and technics, as well, crucially, as a platform for the performance of safety regimens.

The offshore oil industry's commitment to performative safety practices is redundant for reasons additional to those Appel identifies, and sheds light on another crucial difference between ships and platforms. Designed to reassure those enacting them, and the wider public, that dangers can be anticipated and avoided, the rituals ignore the wider existential threat being instigated on the platforms to the planet. The imagined potentially immediate disasters to which the performances respond—blowouts, compromised infrastructure, and so on—are real possibilities, but what of the catastrophe already being perpetuated on timescales not so easily fathomed?

There are fundamental ontological similarities between the roles ships and offshore extractive platforms play in mediating human life at sea. However, platforms also negotiate the fossilized remains of creatures which lived many millennia ago. This necessitates their expanded array of technology and technics, and routinely implicates them in temporal relations far more extensive than those involving ships. As ecocritic Hugo Reinert evocatively reflects, "Fossil is a thing that lived and was buried, and now it keeps a different time."[24] Through their imbrication with petroleum, platforms like *Hibernia* are linked to the deep past. Further, by extracting oil which, when burnt, creates long-lasting carbon emissions, they are conduits to a future in which present actions reverberate for millennia to come. Offshore oil platforms are thus more profoundly implicated in the Anthropocene's temporal horizons than ships. They involve us in mediational contexts with parameters we cannot fully discern. As social historian Dipesh Chakrabarty (2009) suggests, the advent of humans as geological agents poses profound challenges to received understandings of human history, and indeed our comprehension of what it might even mean to be human. The extractive platform is thus the most apposite "archetype of artifice become nature" (Peters, 2015, p. 102) for our present moment because "nature" has now taken on new guises which are increasingly dangerous and incomprehensible to us.

Hibernia operates in the cold reaches of the North Atlantic Ocean far from the literal sight of those who dwell onshore. However, repurposing the media studies concept of the platform helps elucidate important, but often overlooked, ways in which this transiently monumental piece of infrastructure shapes cultures and environments. *Hibernia* has been an invaluable platform for a global oil industry set on profiting from a fragile oceanic environment with a centuries-long history of exploitation. Central to building public consent for petroleum operations, the *Hibernia* project has been presented as in keeping with Newfoundland and Labrador's white settler seafaring traditions, at the same time as it facilitates the province's participation in technological modernity. However, viewing *Hibernia* as itself media, as a platform that makes human ground in a nonhuman element, points us to different, indeed ontological, issues. In this period of ecological crisis and uncertainty, offshore extractive platforms are profoundly unstable ground for being. Through them, we have chosen to yoke ourselves to temporal horizons we cannot fathom, and to cascading consequences that exceed our control.

Notes

1 "Bitumen" from *The Road In Is Not the Same Road Out* copyright © 2015 by Karen Solie. Reproduced with permission from House of Anansi Press Inc., Toronto. www.houseofanansi.com
2 The Grusin, I use nonhuman to refer "variously [… to] animals, affectivity, bodies, organic and geophysical systems, materiality, or technologies" (Grusin, 2015, p. vii), while acknowledging "the human has always coevolved, coexisted, or collaborated with the nonhuman—and that the human is characterized precisely by this indistinction from the nonhuman" (Grusin, 2015, p. x).
3 Pitt Scott (2020) cites Ian Rankin's *Black and Blue* (1997) as an excellent example of offshore noir. For eco-horror, see Amazon Prime series *The Rig* (2023).
4 As Davis and Todd highlight, the concept of the Anthropocene fails "to adequately account for power relations" (Davis and Todd, 2017, p. 763). However, I agree with them that "rather than abandon the term [...] we feel that the Anthropocene betrays itself in its name: in its reassertion of universality" (Davis and Todd, 2017, p. 763).
5 Danine Farquharson and I (2022) refer to this region as the cold water offshore in our book *Cold Water Oil: Offshore Petroleum Cultures.*
6 References to dread do recur in writing by offshore oil workers and their families. In *Rig Wives: Stories*, Newfoundlander Kelly Earle writes: "I worry a lot about my husband's safety going to work, while he is at work, and on his way home from work. Every. Single. Time" (Earle, 2022, p. 7). See also Heffernan (2009).
7 See for instance, https://atlantic.ctvnews.ca/n-l-says-its-waters-are-home-to-low-carbon-oil-scientists-say-that-doesn-t-exist-1.6334256
8 Dwayne Avery discusses the elision of the Husky spill in other contexts in his 2020 *Journal of Environmental Media* article.
9 Oil media take many forms. As environmental humanist Stephanie LeMenager argues in *Living Oil: Petroleum Culture in the American Century*, "compelling oil media are everywhere. Films, books, cars, foods, museums, even town are oil media" (LeMenager, 2014, p. 11).
10 I use the term more-than-human to accentuate the "primacy of relations over entities (including the 'human')" (O'Gorman and Gaynor, 2020, p. 717).
11 Kinder coins "platform energetics" to refer to the "interrelation between the material energy systems and relations on which digital media platforms rely and the energy imaginaries produced and circulated on these platforms" (Kinder, 2023, p. 60). While I do mention data gathering in the context of *Hibernia* below, and, like Kinder, discuss the platform's "soft" and "hard" elements, I am interested in using the platform concept in ways which include but extend beyond digital infrastructures.
12 The island the British called Newfoundland, and the French *Terre-Neuve*, was already home to the Indigenous Beothuk and Mi'kmaq. *Ktaqmkuk* is the Mi'kmaw name for Newfoundland.
13 The Canada-Newfoundland and Labrador Offshore Petroleum Board (C-NLOPB) publishes incident disclosure information on its website, but these reports are not always complete. For instance, "an authorized discharge from the *Hibernia* platform" was reported in July 2020, at which time the "volume and composition" of the discharge were "still being determined." Updated information was never published. I arrived at the estimate above by adding quantities of discharged materials from 1997–2022 listed on the C-NLOPB's website.

14 For more information see Scovil, de Jourdan, and Speers-Roesch (2022).
15 For more on filmic speculations about Newfoundland's oil-related future see Jekanowski (2022).
16 A copy of the sheet is included in the oil industry-related archives of the Centre for Newfoundland Studies at Memorial University.
17 Newfoundland joined Canada in 1949 and has retained a sense of separateness. See Conway (2021).
18 A plethora of data, ranging from oil drilling and production parameters, to meteorology, to ship and helicopter movements, have been gathered at *Hibernia* from the time of its installation on the Grand Banks. See, for instance, Hurwitz (1998) and Hammond (2005).
19 Floating Production Storage and Offloading (FPSO) vessels, such as the Grand Banks' *Terra Nova* FPSO, constitute yet another genre of offshore "ground" blurring boundaries between ship and platform. While technically ships, they do not resemble the vessels John Durham Peters discusses because of their more platform-like features and fossil fuel extractive purposes.
20 For more on distinctions between ships and oil platforms see Polack (2022).
21 That the lore associated with operating offshore extractive platforms has, in contrast to that of Holocene-era seafaring, barely seeped into shore-based cultures may primarily be because of the comparative newness of this technology, its "infrastructural brutalism" (to use Michael Truscello's phrase), and/or its inaccessibility to all except those employed to operate it.
22 Anthropologist Stefan Helmreich makes a similar distinction between ships and platforms on the basis of propulsion (Helmreich, 2021, p. 3).
23 Their sea-based GBSs alone must withstand powerful winds, waves, icebergs, and sea ice. They also need to contain safely sealed chambers and piping to store and transport massive quantities of oil, and resist corrosion and vibration.
24 Eco-horror television series *The Rig* (Macpherson, 2023) posits suppression of just this kind of meditation as cause for nature's revenge. Mark Wahlberg's character in *Deepwater Horizon* (Berg, 2016) also keeps a talismanic fossil.

References

Ad Vantage Productions and G.M. Media. (1998). *Hibernia: A Pioneer's Journey to the Grand Banks* [Film].

Anderson, B.R.O. (1983). *Imagined Communities: Reflections on the Origin and Spread of Nationalism*. Verso.

Appel, H. (2019). *The Licit Life of Capitalism: U.S. Oil in Equatorial Guinea*. Duke University Press.

Atlantic Canada Offshore Petroleum Training and Qualifications Committee. (2020, April). *Atlantic Canada Offshore Petroleum Standard Practice for the Training and Qualifications of Offshore Personnel*. https://www.capp.ca/wp-content/uploads/2020/05/Atlantic-Canada-Offshore-Petroleum-Standard-Practice-for-the-Training-and-Qualifications-of-Offshore-Personnel-369388.pdf.

Avery, D. (2020). Death by a thousand spills: How corporate branding and media strategies downplay the risk of offshore oil spills in Canada. *Journal of Environmental Media*, 1 (2), 225–244.

Barney, D. (2017). Who we are and what we do: Canada as a pipeline nation. In S. Wilson, A. Carlson, and I. Szeman (Eds.), *Petrocultures: Oil, Politics, Culture* (pp. 78–119). McGill-Queens University Press.

Berg, P. (Director). (2016). *Deepwater Horizon* [Film]. Summit Entertainment; Participant; TIK Films; Di Bonaventura Pictures; Closest to the Hole Productions; Leverage Entertainment; Leomus Pictures.

Bird, L. (2020, December 13). Scientists cast doubt on seismic testing environmental mitigations in N.L.'s offshore. CBC News. https://www.cbc.ca/news/canada/newfoundland-labrador/seismic-testing-concerns-offshore-newfoundland-1.5832466.

Burke, C.M., Montevecchi, W.A., and Wiese, F.K. (2012). Inadequate environmental monitoring around offshore oil and gas platforms on the Grand Bank of Eastern Canada: Are risks to marine birds known? *Journal of Environmental Management*, 104, 121–126.

Carter, A.V. (2020). *Fossilized: Environmental Policy in Canada's Petro-Provinces*. UBC Press.

Chakrabarty, D. (2009). The climate of history: Four theses. *Critical Inquiry*, 35(2), 197–222.

Charman, C. (2014). "There are things you don't get over": Resistant mourning in Lisa Moore's *February*. *Studies in Canadian Literature*, 39(2), 126–148.

C-NLOPB (2023, August 1). *Incident Disclosure*. https://www.cnlopb.ca/incidents/.

Context Energy Examined. (2017, November 16). 5 things about hibernia you probably didn't know. https://context.capp.ca/energy-matters/2017/hibernia-5-things/.

Conway, S. (2021). Post-confederation Newfoundland(er) identity, 1949–1991. *Newfoundland and Labrador Studies* 36(2), 217–258.

Davis, H. and Todd, Z. (2017). On the importance of a date, or decolonizing the Anthropocene. *ACME an International E-Journal for Critical Geographies*, 16 (4), 761–780.

Dijck, J. (2022). Fear and Anxiety. In K. Earle (Ed.), *Rig Wives: Stories*. Flanker Press Ltd.

Earle, K. (2022). Fear and Anxiety. In K. Earle (Ed.), *Rig Wives: Stories*. Flanker Press Ltd.

Equinor ASA. (n.d.) *The Bay du Nord Project*. https://www.equinor.com/where-we-are/canada-.bay-du-nord

Fraser, G.S. and Carter, A.V. (2018). Seabird attraction to artificial light in Newfoundland and Labrador's offshore oil fields: Documenting failed regulatory governance. *Ocean Yearbook*, 32(1), 265–282.

Fraser, G.S., and Racine, V. (2016). An evaluation of oil spill responses for offshore oil production projects in Newfoundland and Labrador, Canada: Implications for seabird conservation. *Marine Pollution Bulletin*, 107(1), 36–45.

Geo Centre. (2022). *Oil and Gas Gallery*. St. John's.

Government of Canada Department of Fisheries and Oceans. (2018). *Canada's Ocean's Now: Atlantic Ecosystem 2018*. https://waves-vagues.dfo-mpo.gc.ca/library-bibliotheque/40781987.pdf.

Government of Canada Department of Fisheries and Oceans. (September 1985). *Position Statement on the Hibernia Development Project Prepared for The Hibernia Environmental Assessment Panel*. https://waves-vagues.dfo-mpo.gc.ca/library-bibliotheque/40629107.pdf.

Government of Newfoundland and Labrador Industry, Energy and Technology (n.d.). *Projects*. https://www.gov.nl.ca/iet/energy/petroleum/offshore/projects/.

Grusin, R.A. (2015). Introduction. In R. Grusin (Ed.), *The Nonhuman Turn*. University of Minnesota Press.

Hammond, T.R. (2005). *A Case Study of Hibernia Illustrating the Benefits of AIS Transponders on Oilrigs*. Government of Canada.

Heffernan, M. (2009). *Rig: An Oral History of the Ocean Ranger Disaster*. Breakwater Books.

Helmreich, S. (2021). Flipping the ship: Ocean waves, media orientations, and objectivity at sea. *Media + Environment*, 3(2).

Hibernia Management and Development Company. (2016). *Hibernia's Billionth Barrel: Hibernia: Newfoundland and Labrador's Quiet Workhorse* [Film] https://youtu.be/x66K-734CsQ.

Hibernia Oil and Gas Field Project, Newfoundland and Labrador. (2021, May 21). *Offshore Technology*. https://www.offshore-technology.com/projects/hibernia-oil-gas-field-project/?cf-view.

Hurwitz, L.D. (1998). Connecting Hibernia. *Hart's Petroleum Engineer International*, 71(5), 31–34.

Hibernia: A Pioneer's Journey to the Grand Banks. (1998). [Film]. Ad Vantage Productions and G.M. Media Productions. St. John's.

Jekanowski, R. (2022). Contested and emergent futures: Film and energy regimes of the Newfoundland offshore. In F. Polack and D. Farquharson (Eds.), *Cold Water Oil: Offshore Petroleum Cultures* (pp. 116–136). Routledge.

Jørgensen, D. (2014). Mixing oil and water: Naturalizing offshore oil platforms in American aquariums. In R. Barrett and D. Worden (Eds.), *Oil Culture* (pp. 267–288). University of Minnesota Press.

Kinder, J., Pasek, A., Lin, C.K., and Cooper, Z.G.T. (2023). *Digital Energetics*. University of Minnesota Press.

Kinder, J., and Stepanik, L. (2020). Editorial introduction | oil and media, oil as media: mediating petrocultures then and now. *MediaTropes*, 7(2), i–xvi.

LeMenager, S. (2014). *Living Oil: Petroleum Culture in the American Century*. Oxford University Press.

Mambra, S. (2023, March 9). 5 biggest oil platforms in the world. Marine Insight. https://www.marineinsight.com/offshore/5-biggest-oil-platforms-in-the-world/.

Maynard, L. (Ed.) (1997). *Hibernia: Promise of Rock & Sea*. Breakwater Books.

MacLeod, P. (Director) (1981). *Offshore Oil: Are We Ready?* [Film]. ONF; NFB.

Macpherson, D. (Creator) (2023). *The Rig*. [Television Series]. Amazon Prime.

Mobil Oil Canada Ltd. (1985). *Hibernia Development Project, Environmental Impact Statement*, Volume 1 Summary.

Murphy, J. (1912). *Old Songs of Newfoundland*. Memorial University Digital Archives Initiative.

O'Gorman, E. and Gaynor, A. (2020). More-than-human histories. *Environmental History*, 25(4), 711–735.

Ostle, C., Thompson, R.C., Broughton, D., Gregory, L., Wootton, M., and Johns, D. G. (2019). The rise in ocean plastics evidenced from a 60-year time series. *Nature Communications*, 10(1), 1622–1622.

Peters, J.D. (2015). *The Marvelous Clouds: Toward a Philosophy of Elemental Media*. University of Chicago Press.

Pitt Scott, H. (2020). Offshore mysteries, narrative infrastructure: Oil, noir, and the world-ocean. *Humanities*, 9(3), 1–15. doi:10.3390/h9030071.

"Platform." (2010). In *Oxford Dictionary of English* (Third Edition). Angus Stevenson (Ed.). Oxford University Press.

Poell, T. (2020). Three challenges for media studies in the age of platforms. *Television & New Media*, 21(6), 650–657.

Polack, F. (2022). Encountering the nonhuman in North Atlantic oil catastrophes. In F. Polack and D. Farquharson (Eds.), *Cold Water Oil: Offshore Petroleum Cultures* (pp. 40–60). Routledge.

Polack, F. and Farquharson, D. (Eds.) (2022). *Cold Water Oil: Offshore Petroleum Cultures.* Routledge.

Porter, S. (Ed.) (2009). *Random Thoughts: Memories & Writings of Greg Tiller.* DRC Publishers.

Rankin, I. (2016). *Black and Blue.* Orion.

Roberts, T. (2016, June 9). A crown jewel for ExxonMobil: Hibernia a quiet workhorse in N.L.'s offshore. CBC News. https://www.cbc.ca/news/canada/newfoundland-labrador/hibernia-crown-jewell-1.3622103.

Reinert, H. (2022). Fossil. *Environmental Humanities*, 14(3), 575–578.

Royal Commission on the Ocean Ranger Marine Disaster. (1984). *Report One: The Loss of the Semisubmersible Drill Rig* Ocean Ranger *and its Crew.* Canadian Government Publishing Centre.

Scovil, A.M., de Jourdan, B.P., and Speers-Roesch, B. (2022). Intraspecific variation in the sublethal effects of physically and chemically dispersed crude oil on early life stages of Atlantic cod (Gadus morhua). *Environmental Toxicology and Chemistry*, 41(8), 1967–1976.

Sekula, A. (2002). Between the net and the deep blue sea (rethinking the traffic in photographs). *October*, 102, 3–34.

Smellie, S. (2021, April 19). After almost 3 decades, cod are still not back off N.L. scientists worry it may never happen. CBC News. https://www.cbc.ca/news/canada/newfoundland-labrador/cod-return-1.5992916.

Solie, K. (2015). *The Road in Is Not the Same Road Out.* Anansi.

Srnicek, N. (2016). *Platform Capitalism.* Wiley.

The Canadian Press. (2023, March 29). N.L. says its waters are home to 'low-carbon' oil. Scientists say that doesn't exist. CTV News. https://atlantic.ctvnews.ca/n-l-says-its-waters-are-home-to-low-carbon-oil-scientists-say-that-doesn-t-exist-1.6334256.

Travis, C., Ludlow, F., Matthews, A., Lougheed, K., Rankin, K., Allaire, B. et al. (2020). Inventing the Grand Banks: A deep chart: Humanities GIS, Cartesian, and literary perceptions of the north-west Atlantic fishery ca 1500–1800. *Geo: Geography and Environment*, 7(1), 1–22.

Truscello, M. (2020). *Infrastructural Brutalism: Art and the Necropolitics of Infrastructure.* MIT Press.

Urry, J. (2014). *Offshoring*. Polity Press.

Van Dijck, J., Poell, T., and De Waal, M. (2018). *The Platform Society: Public Values in a Connective World.* Oxford University Press.

Wilson, R., and Fazel, B. (2022). Fire rated enclosures: A case study. *ieee transactions on industry applications*, 58(5), 5675–5686.

Wood, D. (1999, February). "Hibernia." *En Route*, pp. 48–58. Air Canada.

7 Charting Courses

AIS and Navigational Images in the High North Atlantic

Asko Lehmuskallio

Introduction

Navigation, following its etymology of 'sailing,' is the art and skill of relating to a changing and moving environment (cf. Vigh, 2009). For centuries, before steam and oil altered the role of wind in maritime travel, sailors relied on specific moments when the force of winds could be harnessed for navigation. Sailing vessels needed to wait patiently for opportune times when wind patterns aligned favourably.

However, seafaring was not a matter of convenience. It demanded meticulous attention to environmental conditions. Sailors considered the wind's direction, strength, and predictability. Storms, fogs, and extreme cold posed significant challenges. Navigators divined future weather conditions to ensure safe passage. Even today, environmental awareness remains crucial, as the interplay of wind, currents, and weather profoundly impacts seafaring.

Going by sail on the High North Atlantic continues to call for a focus on ways of relating to the winds, adjusting sails in relation to their force and direction, in temperatures that remain bearable. At the same time, navigational techniques for knowing where one is, and which direction one is heading towards, have multiplied. Airplanes, helicopters, drones, and satellites aid today seafaring, providing a vertical geopolitics (cf. Elden, 2013) from above in addition to the one focusing on the depths and the surface level, hence expanding our understanding of navigating the ocean.

Since the 1990s, the Automated Identification System (AIS) has been developed and implemented for maritime use, allowing vessels to identify themselves and track each other's movements via VHF (Very High Frequency). AIS transponders transmit radio waves on a very high frequency rate, sent through an antenna. The radio waves emitted indicate a unique identification number (MMSI), position (via GPS), direction (via a gyrocompass) and rate of turn (via an indicator), and these can be decoded with a receiver.[1] AIS information can be accessed also remotely on websites, as seen in Figure 7.1.

AIS provides additional cues for navigational purposes, and has become a staple for maritime navigation, including on sailing boats, especially those

DOI: 10.4324/9781003478805-7

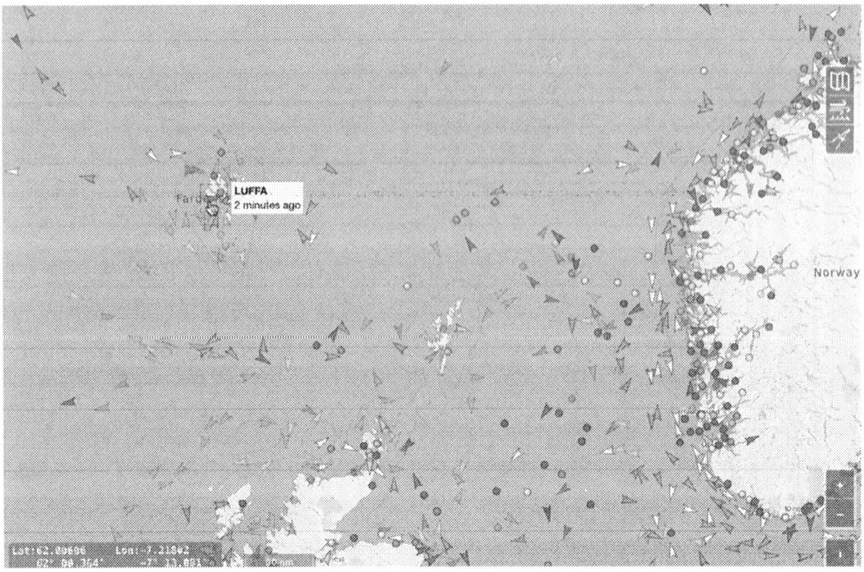

Figure 7.1 The sailing vessel "Luffa" has automatically provided its position via an AIS transponder two minutes ago and is visible as a symbol on the site VesselFinder, a freely available AIS ship tracker web site
Source: Screenshot.

traversing open waters. The High North Atlantic plays a specific role for AIS, as Norway has significantly advanced the development and implementation of these technologies. The Norwegian Coastal Administration (*Kystverket*) has established AIS Norway, a network for receiving AIS signals from vessels along the Norwegian coastline, and the Norwegian Space Agency (*Norsk Romsenter*) and Defence Research Establishment (*Forsvarets forskningsinstitutt*) have been important in developing the NorSat-satellites in collaboration with University of Toronto Institute for Aerospace Studies' Space Flight Laboratories (UTIAS/SFL). The NorSat-satellites receive AIS signals from below, providing for the visualization of a more comprehensive maritime picture than would be possible by relying only on surface-level AIS signals.[2]

In the following I will focus on the example of going by sail in order to pinpoint modes of navigational knowledge that are needed to move with a sailing vessel within the High North Atlantic. I will focus on embodied forms of navigation when going by sail, the role of AIS among these, as well as on the importance of *kairos* for navigating in uncertain conditions. Accidents play a significant role in the development of navigational aids, and I will provide select examples in order to highlight their role in the further development of tools and techniques for charting courses on sea. I suggest that navigation significantly relies on navigational images, that is, on specific temporally unfolding means of interaction within the environment in order to know where one is, and where to move towards. By focusing on select examples of navigating the seas, I hope also to show how navigating oceans provide a way for rethinking the role that environments provide for our understanding of media (cf. Peters, 2020).

Navigational Images

During navigation specific cues in maps, charts, plotters as well as in the immediate environment are used to form navigational images which orient situational interaction. These images emerge in embodied interactions with the environment in order to situate oneself and/or to know where to move next. They are not mainly mental representations in the heads of observers, nor pictures seen on visual media, but emerge as an in-between as humans interact within environments. I understand them as specific kinds of images that evolve during the temporally unfolding act of looking. As Hans Belting (2007, translation AL) describes it: "[Images] cannot be located only 'there,' on a canvas or in a photo; neither are they located only 'here,' in the head of the beholder. The look constitutes the images in the range between 'here' and 'there'." Navigational images, from this perspective, emerge through modes of interaction, in order to situate ourselves as well as to direct our energies towards a future location.

Conceptually, the understanding of navigational images follows a suggestion for a broader anthropology of the image, which is interested in the interrelations between images, bodies and media in particular environments (Belting, 2011). The suggestion of environmental images by Kevin Lynch (1960) is particularly useful in developing the notion of navigational images, but whereas for Lynch environmental images are understood to be mental representations, my understanding follows Hans Belting's and Tim Ingold's (2002) critique of a strict division between external and internal representations, suggesting that images emerge as an in-between during interaction.

From this perspective, maps and charts provide visual cues for navigation, and are part of the arsenal used in situ and actu for navigational imagery. They are resources that can be drawn upon in the formation of situational navigational imagery.[3] As work in critical cartography has highlighted, the interaction of politics and power in map creation needs to be recognized, maps are not neutral data visualizations (Kitchin and Dodge, 2007). Rather, maps are social artifacts that need to be interpreted as authorial documents rather than mere representations of data (Koch, 2004). Importantly, work in critical cartography has questioned an ontological security of maps, focusing on the iterative processes through which mappings are constantly remade (Rose-Redwood, 2015). The narratives and contexts behind maps are highlighted, moving beyond the mere representation of geographical features (Caquard and Cartwright, 2014). Furthermore, critical cartography increasingly pays attention to the role of maps in shaping our understanding of the world (Kitchin and Dodge, 2007). Maps and nautical charts are technologies of power that can influence perceptions of reality, and hence the situational 'fit' of navigational practices (Bauer, 2009). As maps and charts are cultural artifacts that shape our perceptions of the world, they do this shaping by providing resources for navigational imagery to emerge.

In addition to the concise information available on maps and charts, navigators also rely on environmental cues, such as the direction and height of waves, forms of coastlines, cairns and lighthouses, thus fusing natural and manmade navigational aids when deciding how to navigate. Navigational images emerge in the interaction within specific environments using navigational aids that are situationally available. As will be shown later, these resources rely on various forms of movement: that of ships and sailors, those of repeating patterns in the environment, movement within technical devices themselves, as well as the movement of these technical devices in environments.

The notion of navigational images allows us to focus on the importance of interaction in image use, on the 'work' that images do in our lives. It describes a very specific mode of imagery, that which we use in order to locate ourselves and which we use in order to decide where to move towards. At the same time it questions the usefulness of a strict categorization of imagery based on the visual media within which they appear (e.g. as printed photographs, paper maps, or plotted charts), understanding the latter rather as resources that we use for images to emerge.

Sailing

Sailing, as an exercise in relating towards the winds, is primarily concerned with the here-and-now, as well as with ideas and models of what the near future will bring forth. The here-and-now is evaluated while on the move, by observing one's surroundings on the ship's deck. Sailors feel the vessel's tilt and rhythm, regularly adjusting their bodies in a literally moving environment, and monitor sensor readings on deck providing information on wind direction, speed of the vessel as well as water depth below it. When considering the future, they rely on calculations and visual representations that must align with the underlying models used during their creation.[4]

Those who have ever sailed know that winds do not blow in just one linear way, but keep moving, changing direction and force, so that sailors need regularly to adjust how they relate to the changing winds. Sudden gusts call for quick reactions, and close to islands the winds may behave unpredictably, so that constant sensory vigilance is necessary. Navigational images constructed in the moment of sailing allow for more precise adjustments in relation to the winds, adjustments which are always reaffirmed affectively in an embodied way as the movements of one's body and the feel of the air provide a sense of how to relate.

Nautical charts visible on plotter screens supplement embodied forms of sensing and knowing, providing particular condensations of models of one's position within an environment. They provide navigational cues, making available in visual form "an environment augmented (and temporalized) by fully or partially scoped components – in which we find ourselves in one another's and the scopic components' response presence, without needing to be in one another's physical presence" (Knorr Cetina, 2014, p. 47). These nautical charts relay a range of information from various sources, assisting in locating one's self in a specific spatiotemporal setting. They are also used in order to facilitate decision-making on where to go next, in which case their use is mainly future oriented.

The screen showing nautical charts acts as a scopic medium, "visually present[ing] and project[ing] events, phenomena, and actors that would otherwise be separated by distance and would not be visible from a single standpoint" (Knorr Cetina, 2014, p. 43). Augmenting and expanding local situations, these scopic media allow for organising events from afar and from different temporalities. They are used both as locative media (Thielmann, 2010), to know where one is, and are anticipatory, oriented towards the future. They are aggregated in order to imagine futures that become socially effective, calling for response presence (Knorr Cetina, 2009). Scopic media, providing resources for navigational images, used in situ and actu, urge for embodied reactions, which increasingly can also be delegated to (semi-)automated technologies. For example, plotters used on sailing boats can be pre-programmed to follow specific routes, so that the plotter commands how the rudder is automatically moved. They may also just indicate, based on wind and weather models, when a human operator should tack the boat, that is, conduct a maneuver by turning the bow of the sailing boat through the wind to change direction and adjust the sheets of the sails to properly trim them on the new tack.

Depending on the amount of information visible on a screen, navigational images, in sailing, might rely on sensor readings conducted on a moving vessel (e.g. depth, wind, location, speed, temperature, air pressure, etc.), as well as on a range of information collected from afar, such as previously gathered satellite imagery, stitched together in order to provide a model and 'bird's-eye-view' of one's immediate environment. Also, information of vessels nearby may be collected via the automated identification system AIS as well as one's own location triangulated with the help of GPS (or Glonass, Galileo or BeiDou).

Navigational images thus complicate how we might think about forms of representation and their relation to the act of recording.[5] In contrast to operative images, which are usually considered to be invisual and nonhuman, following particular algorithmically prescribed paths (Farocki, 2004; Feiersinger, Friedrich, and Queisner, 2018; Parikka, 2023), the navigational stresses the volatility and opacity of the situations within which one is to move. It provides thus a perspective that is specifically interested in the means that people use for navigating environments (cf. Verhoeff, 2012 discussing a visual regime for navigation), even if some of these means are automated as part of operative chains.

Expanding *Kairos* with Navigational Techniques

Kairos, the propitious moment, is of particular importance for sailing. Without winds, with too strong winds, or winds blowing from directions that are too difficult to sail, navigation becomes a futile activity. Most seafarers have encountered sudden changes in weather conditions, the difficulty of following one's way in storms, limited vision as night falls, or fog making it impossible to see any other ships or coastlines. Since kairos does not always prevail, accidents are bound to happen. Paul Virilio has famously stated, that "[t]o invent the sailing ship or steamer is *to invent the shipwreck*" (Virilio, 2007, p. 10, orig. emphasis). Shipwrecks, early on, have called for better and more meaningful ways of knowing where one currently is, as well as for divining which direction to go towards.

Colonial and imperial phantasies of riches and domination, of exploration and fame have been for centuries coupled with difficulties of knowing one's location on sea. While Cristoforo Colombo was able to use latitude measurements while sailing towards America, he could not mark his longitudinal position on his route, and famously thought to have landed somewhere very different from where he actually was. Nevertheless, he was able to return showcasing a means to find a new route to fulfil imperial phantasies and was made "admiral of the ocean." "Successes" such as his, coupled with the difficulties of knowing one's longitude on sea, provided impetus for European rulers from Spain, Portugal, France, and England to call out prizes for finding means to know one's position on sea.

Importantly, a disastrous shipwreck in 1707 at the rocks west of Scilly influenced the passing of a Longitude Act in 1714, offering a substantial reward for anyone who could provide a means for conducting reliable longitude measurements on sea adding to a heightened interest for finding solutions. As is well known, decades later two different methods started to provide relatively reliable results, one based on the lunar-distance method, and another one based on the development of a reliable marine chronometer that could be set to a specific time (e.g. the one measured at the Royal Observatory), and carried along onboard and compared to noon by studying the position of the sun vis-à-vis the horizon. Because marine chronometers were expensive and could not always be relied on, nautical charts for calculating longitude based on the lunar-distance method were regularly used for cross-verification purposes.[6]

The ability to calculate longitude on sea expanded the range and possibilities for seafaring significantly, but perhaps even as importantly, allowed for novel means for mapping the oceans. Wide areas became representable on cartographic projections as ships collected data points with relevant environmental information from their voyages, "in order to achieve the miracle of reference" (November, Camacho-Hübner, and Latour, 2010). The possibility to know one's location in space (often with the help of marking time, as e.g. when using marine chronometers for doing so) provided simultaneously a means to visually represent space. Navigation, and the representations used for navigation, are tightly interwoven.

The Example of AIS

Nautical charts, visible on screens when on the move, mark a range of information that is of importance when sailing. These scopic media are never encountered as still stable representations that may be studied at length, but are during sailing usually only glanced upon as the boat keeps rhythmically rocking with the waves, and the hull is tilted in response to the wind's pressure on the sails. The navigational image, formed during the interaction with charts on a boat that is itself on the move, does not take everything that is displayed on these nautical charts into account, but only bits and pieces that are relevant for a specific situation (e.g. when trying to avoid rocks on port tack). The winds, currents, the tide, and the amount of water below a ship's keel affect also how the ocean itself moves, so that wind and water together form the elemental media through which navigation as an activity within a moving environment occurs. This makes navigation a volatile activity, as those who move do so in an environment that itself keeps changing, needing to focus on various resources that are useful for navigation.

In addition to the changes in one's own movement as a ship traverses the seas, as well as the movements of the air and water through which one navigates, also the navigational images, as future-oriented representations, may change as information on a display is updated. In contrast to more traditional stable visual representations, in electronic nautical charts used onboard specific details change over time, such as the exact location of a vessel on a represented map, or the strength of the wind depicted as vectors. These changes of specific details indicate a call for response presence, that is, of information to be acted upon as deemed necessary, and hence have an impact on the navigational images formed during situated interactions.

One of the areas that remains difficult to trace beforehand, and needs to be conducted in situ and actu, is knowledge of other vessels close-by that might impact one's route. For sailors, for example tankers and other cargo ships are important to locate early on, as they need to be given way and might take up the whole breadth of a shipping channel. Early on, radar technology was developed in order to enhance situational awareness of other vessels, beings and objects in the air and sea, with some scholars tracing the history of the digital display so important for scopic media to the military use of radar technologies in the 1940s (Thielmann, 2018). Radio waves have continued to provide for important information for sailing purposes, not least via the implementation of a VHF system for marine radio used to call port authorities, report accidents, contact other vessels in one's vicinity, and so on.[7]

Instead of 'just' sensing the environment, as done for example with radar, boats are also increasingly equipped with devices that explicitly emit signals for others to trace. Automated identification systems (AIS) are mandatory for a range of large commercial vessels on sea, but increasingly also sailors on sailing boats use them, particularly when sailing in possibly harsh conditions, such as during stormy weather which is frequent particularly from September to April in the High North Atlantic. Although the history of AIS is complex (cf. Pezzani and Heller, 2019, p. 886), efforts to enhance both vessel tracking capabilities as well as the situational awareness of navigators is particularly tied to one of the largest environmental disasters in US history, the Exxon Valdez oil spill in Prince William Sound, Alaska. In response to the spill of 11 million gallons of oil, an oil pollution act was passed, which included a call for developing a vessel tracking system for tankers going to Alaska.[8]

AIS, developed for purposes of tracking and navigation on sea, is thus an additional example of the role that accidents play for development of marine technologies.[9] The International Telecommunications Union had recommended in 1998 technical characteristics of a universal shipborne AIS system and the International Maritime Organization mandated the use of this equipment for a range of ships from 2002 onwards.[10] It is considered to be a significant advancement in navigation safety for mariners, comparable to the impact of radar and GPS. Operating within the Very High Frequency (VHF) maritime band, AIS serves multiple purposes: ship identification, target tracking, search and rescue support, streamlined information exchange, and enhanced situational awareness. Due to the environmental disaster emerging from the Exxon Valdez oil spill, AIS was originally designed as a collision avoidance tool. It enables commercial and larger recreational vessels to better perceive each other under various conditions, providing helmsmen with crucial environmental information. By continuously transmitting a vessel's identity, position, speed, and course, AIS facilitates communication among equipped vessels within range. It is also an important part of safety and rescue operations, as sailors on open seas frequently carry AIS devices which are activated if they should fall over board. With the aid of specific shore stations and satellites equipped with AIS receivers, it empowers port authorities and maritime safety agencies to manage traffic and mitigate navigation hazards.

A wide range of additional navigational techniques could be listed, but more important is the insight that navigation brings to our understandings of mediation, which is never perfect, but continues at times to 'miss' the complexity of the world. Navigation, as a volatile activity, is conducted with techniques and technologies that may fail, or that might not be suited to the exact situation and its circumstances. The scopic media, providing resources for navigational images, may scope information that is not correct, or useful at all during a specific moment. Navigation, as a form of movement in an environment that itself moves, is hence never 'perfect' but continues to be equipped with ever more techniques and technologies.

The information sent out by different vessels via AIS is received onboard of ships with purpose-built receivers, but also pooled together by specific land-based stations. This information can be arranged on a cartographic grid on a screen onboard in order to provide those navigating information of vessels around them, mainly their location, direction, and speed. AIS hence extends and enhances embodied practices of lookout, such as the use of sight and of sound (including devices such as for example horns and whistles), which have been a staple of navigational practices since the early days of wayfinding.

While the horizontal range of AIS is limited to ca. 40 nautical miles (ca. 74 kilometres), depending on environmental features and weather, researchers noted in the 2000s that the vertical range is much higher, possibly reaching up to the orbit of the International Space Station orbiting at ca. 400 kilometres. The European Space Agency contacted the Norwegian Defence Research Establishment and provided them with the possibility to install an AIS receiver in order to test space-based AIS (Figure 7.2).

While there was insecurity regarding the feasibility of space-based AIS, the system was switched on on June 1, 2010, and those involved explain how "suddenly we could see a whole world of ship data," allowing to observe about 30 000 ships on the open ocean per day.[11] They point to the significant shifts in synthetic situations as scopic media and the particular constellations they rely on change. Satellite receivers have continued to be further developed, and for example in July 2017 NorSat-1 and NorSat-2 microsatellites were launched with AIS receivers developed by the Trondheim (Norway) based Kongsberg Seatex onboard allowing for more focused tracking of vessels especially also in the High North Atlantic, with the latest micro satellite launched in April 2023.

The AIS receiver installed for testing space-based AIS in 2010 was an additional element to an already established International Space Station, and only the later NorSat microsatellites were launched to orbit in order to focus specifically (although not solely) on facilitating AIS. The question of how to know if and what kinds of ships will be on one's route started to be solved with the help of AIS receivers orbiting the Earth hundreds of kilometres above the ships sending these signals out.

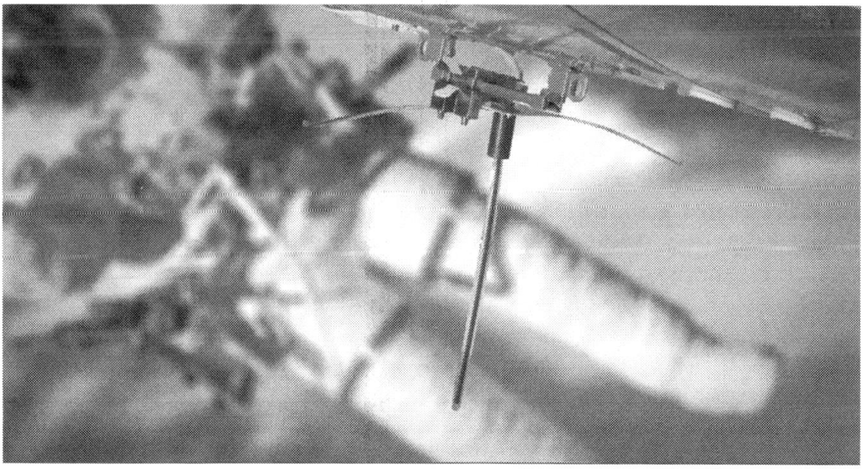

Figure 7.2 AIS receiver installed onboard the International Space Station in order to test the feasibility of satellite-based AIS

Source: Screenshot from NASA film ISS Benefits for Humanity: Found at Sea.

The ability to determine longitude in order to know one's position on sea had an important impact on the resolution and accuracy of maps of the seas, and earlier satellite systems had enabled locating one's own position on a virtual model of the globe. Space-based AIS provided a global view of all those ships emitting information via transponders, again changing what elements in nautical charts could be made visible. AIS thus provided additional complexity to situated navigational imagery, as it allowed for expanded forms of vessel tracking and collision avoidance which at times could cover the whole Earth.

Navigating 'in situ' and 'in actu': The Bifocality of Vision

Scopic media, organizing what becomes visible on screens, act as means to create global totalities, which by showing "it all," fail to show the richness of detail that might be visible otherwise. Describing the experience of seeing the AIS receiver on the International Space Station in action, feeding globally transmitted AIS data to monitoring screens, one of those involved explains: "It was like a blind man starting to see" (NASA, 2014).

This sudden affective experience of an ability to see is accompanied by the paradox that as "the general becomes clear through representation, … the immediate is subject to the fragmenting effects of our limited experience" (Peters, 1997). Describing synthetic situations through the lens of a "bifocality of vision," John Durham Peters follows Benjamin Franklin's reflections on the importance of different kinds of spectacles in order to see close-by (in Franklin's case, the food one eats), and somewhat further (the faces and features of those on the other side of the table in order to understand what they say). Thus detaching synthetic situations from a strict distinction between a "naked, unmediated face-to-face situation" (Knorr Cetina, 2014, p. 60) and one in which "electronic media and screens reconfigure the situation" (Knorr Cetina, 2014, p. 45), a focus on the bifocality of vision allows us to ponder different forms of navigational images and the kinds of assemblages and constellations they rely on, without limiting our focus mainly on either electronics or screens.

The forms of mediation that are particularly important for a specific practice, such as for navigation on sea, are focusing media, such as a port map, nautical chart or weather information relayed via radio. While mediation in general reconfigures and transforms "whole practices, ecologies of relations, knowledges, senses and bodies" (Grasseni and Gieser, 2019, p. 10), there tend to be some forms of mediation that are more relevant to a specific practice than to another. These are "so central that much of the learning and enskillment processes revolve around them. It is with them that the powers of reconfiguration and transformation become most apparent" (Grasseni and Gieser, 2019, p. 10).

During navigation on sea, navigational aids act as focusing media, used within interactions in order to create situational navigational images. AIS information, pooled on a screen, layered on top of other information on nautical charts, is precisely this kind of information that can be used for situational awareness. As an emphasis on bifocality of vision suggests, AIS adds to the resources that are available for navigational purposes on sea, but it does not make earlier forms of navigation useless. Navigational images hence may be supplemented by information provided by AIS, but do not always need to be. So if a specific technology, such as AIS, only partially impacts the means of navigation on sea, what would more elemental cultural techniques (Siegert, 2023) be that make a sustained difference?

The Role of Movement

Following the phenomenological observation of a bifocality of vision mediating our everyday activities, it is clear when sailing, and observing sailors, that navigation on sea does not occur just by focusing on AIS information or even full nautical charts on an electronic screen (or a paper-based map), nor solely by observing one's immediate surroundings.

Rather, navigation on sea, particularly in sailing as an activity of relating to the winds, combines four kinds of modes of knowing, which all rely on different kinds of body movement, ways of observation, analysis, and representation. All of these modes of knowing how to navigate a particular environment may also be thought of as specific ways of data collection, allowing for different kinds of navigational images of "how to" navigate to emerge. These elemental cultural techniques are, first, embodied movement from A to B; second, observing by holding still; third, observing movement within a technical apparatus; and fourth, observing moving devices placed in environments. All of these techniques couple specific forms of movement with particular ideas concerning time and temporalities. The knowledge gained is then available for transcription and transduction, and may become an important resource for situational navigational images. Importantly, although these distinct modes have arguably been developed with successive technological developments and ways of calculating, or estimating, specific connections among data points collected, they all continue to be used regularly in actu and in situ while sailing, also in the High North Atlantic. For example AIS transponders transmit in addition to identificatory information (i.e. the unique identification number MMSI) sensor readings which respond to particular forms of movement, as well as receives this kind of information from other vessels, specific forms of movement being key for AIS to work at all.

Moving from A to B

The actual movement from A to B continues to be the pinnacle of navigation on sea, and happens during phenomenal time. Here the human body needs to physically move within the moving environment, being part of a vessel, and the movement follows meanderings along environmental features to dedicated sighting positions. In order to represent one's experiences in a meaningful way, parts of the environment have to be considered to remain relatively stable, so that they can serve as reference points that may be mediated further. This information can be conveyed with the help of mnemotechniques, using talk, text, and drawings in order to do so, and is today visualized on nautical charts. Sailing instructions, told others, or written down, as well as portolan charts and local pilots boarding a ship, are historical and current examples of this kind of navigational knowledge, in which the seeing and sensing for purposes of navigation is done while being on the move. For centuries, this was the main way of knowing where to sail towards, and tended to be coupled with knowledge of the opportune moments for doing so, so that wind and weather would not make the movement too dangerous, or impossible to do. On the downside, this kind of knowledge is like the social scientist's participant observation: slow to gather, difficult to update, and sightings and measurements need usually to be done when one is able to sense the environment. For purposes of navigation, during daytime, with good enough visibility, if focusing on visible reference points, and during nighttime, if observing moons, planets, and the constellations of stars which serve a similar purpose.[12] This actual movement from A to B continues to be today the raison d'être for AIS systems, as they emit information concerning this movement for others to follow.

Observing by Holding Still

Embodied movement from A to B is lived along lines of becoming (Ingold 2016), instead of through straight vector-like lines. It is a movement that occurs within a moving environment, the elemental media of air and water, a movement that has to take the movements of both kinds of media into account. The sails are adjusted according to the winds, and the hull has to be aligned so that it can bear the swell and changes in directions of waves, particularly as they can become several metres high on the High North Atlantic. This embodied movement provides specific ways of knowing how-to navigate, both in the moment, as well as through mnemotechniques that help to trace past forms of movement.

As a counterpoint to the importance of embodied movement in order to know how to navigate, the cessation of movement, the ability to hold still, provides for an important mode for collecting information in order to know how-to navigate. For centuries, location work has been conducted by observing the rotating 'heavenly clock' vis-à-vis one's own position, in purpose-built observatories, by holding still long enough so that the moving celestial bodies could be observed and marked on charts or maps in order to tell one's whereabouts, as well as the local time. In doing so, virtual models of the relationships between the Earth and heavenly bodies are of importance when situating one's self within a framework, as done, for example when calculating one's position based on an orbit containing latitude and longitude. The associated temporality is circular and based on repetition, for example daily (e.g. dawn, noon, dusk), monthly (e.g. lunar calendar), yearly (e.g. seasons and sun), or in a pair of pairs pattern (e.g. the Transit of Venus in front of the sun repeats in pairs every 243 years).

Using the sundial, or assessing latitude with the help of a quadrant or sextant are examples for this mode of location work.[13] Constructing a technical apparatus for the purposes of standing still in order to observe, measure, and record moving heavenly bodies was considered to be particularly useful for finding a means to locate oneself also on waters far away from home, as well as for centralizing and standardising measurements of time. Especially lunar movements were considered to be promising for such an endeavour, and a range of scholars tried long and arduously to gather enough data in order to be able to calculate a way for measuring longitude on sea, as discussed earlier. This information collected by observing still is a relevant basis for the development of models that are used for navigation (e.g. the latitude/longitude -system), and is activated in the form of a framework (or grid) within which one can position oneself. It provides spatial structure to the situational navigational images formed during interaction.

Observing Movements within a Technical Apparatus

If the hitherto described modes relied on 1) one's body moving through the environment, or 2) on being fixed and observing the moving surroundings, a third mode of navigation came about as the invention of a stable enough marine chronometer (by John Harrison) allowed sailors to carry a technical apparatus, and hence to "carry time" with them. Time did not need to be read anymore (solely) from the environment, by for example calculating the sun's angle, but one could focus on observing the movements within a technical device. This constellation is based on a model of how particular materials interact over time, and what is of interest here is the movement within a material medium itself: in the case of Harrison's clock, the pointers of the clock in relation to the dial.

For marine navigation, the invention of the marine chronometer (along with the lunar-distance method which Nevil Maskelyne made actionable around the same time with the help of the nautical almanac) provided the ability to calculate longitude on sea, which allowed both to locate oneself on a virtual grid known since Antiquity, but also to create ever more accurate representations of the sea, leading to a blooming and blossoming of hydrographic societies, which provided since the late 18th century contemporary hydrographic big data in the form of charts and maps for knowing the seas.

The chronometers were still synchronised based on celestial observations such as those by the Royal Astronomer in Greenwich, who provided a time ball mechanism at noon so that seamen could correct their chronometers onsite.[14]

AIS includes a gyrocompass, which similarly to Harrison's marine chronometer contains movement within the technical device itself, being important for measuring direction. While location is assessed using GPS with the help of the movement of satellites orbiting the Earth, in a gyrocompass the movement is embedded within the form of a spinning wheel or rotor, allowing for observations based on movements within a technical apparatus. In this third case, the observations are standardized by focusing on movements within a technical apparatus, instead of those in physical environments. The apparatus becomes the site for strict repetition that may be relied upon in order to form even tighter structure into situational navigational images.

Observing Moving Devices Placed in Environments

Increasingly, technological devices are placed within natural environments, such as lighthouses on islands or buoys attached to the ocean floor aiding location work on sea with the help of the signals they emit. Of interest for navigational images such as those including AIS are satellites, orbiting around the Earth, which carry their own time (in atomic clocks) that is used in order to provide correct calculations of receivers' locations. In the 1960s, TRANSIT, an early satellite navigation system was implemented to provide location information to US ballistic missile submarines, as well as for purposes of marine surveying (Figure 7.3). Five satellites orbited the Earth every two hours, and submariners could measure their position by identifying the changing frequencies of each satellite's radio signal.

Figure 7.3 A receiver above the Prime Meridian at the Royal Observatory in Greenwich, UK, to calibrate TRANSIT, the first satellite navigation system launched. It was launched in order to provide accurate location information to US Navy Polaris ballistic missile submarine.

Source: Photographer: unknown, 1964.National Maritime Museum, Greenwich, London.

In the case of positional information conveyed via GPS, the spatiotemporal location of a sailing boat is assessed by relying on a constellation of satellites orbiting the Earth. These satellites broadcast signals, which can be received with a purpose-built GPS receiver, and the receiver has an algorithm embedded to calculate its three-dimensional position (latitude, longitude and altitude). The signal emitting satellites, which orbit the Earth, serve as reference points, allowing to position one's self in a virtual orbit that contains a grid allowing for positioning according to latitude, longitude, and altitude. Here location is assessed with the help of technical devices moving in an orbit around the Earth, and this is conveyed via AIS systems to other vessels, base stations on land, as well as satellites orbiting the Earth.

The Earth is turned into a laboratory, and observations "in the wild" may be conducted based on standardized apparatuses that carry time, tightening the grid, net, and structure that may be used for navigational images. We are provided with ever tighter virtual models within which to navigate, allowing for higher resolution. Within this development, AIS remains an interesting example, closely tied to a broader interest in the Internet of Things, a vision and example of a high-resolution global meshwork of signal-receiving and -emitting devices placed in environments.

Conclusions

Navigational images are used in order to locate oneself (or one's vessel) in the "here and now," as well as in order to direct oneself towards a future direction. They rely on nautical charts, maps, and different kinds of environmental cues. The work of locating one's vessel is conducted in relation to a reference system that is increasingly detached solely from embodied observation onsite, referring rather to virtual models facilitating action, such as the longitude/latitude model or the tightening grid for location work that AIS allows for. In order to locate oneself, one has to move within a specific constellation that allows for positioning within it. This positioning needs points of reference, which become inscribed within representations used for locational purposes (e.g. in portolan charts; gridded maps incl. longitude and latitude; AIS vectors; or satellite images).

Over time, constellations used for positioning oneself change, and hence also the ways in which specific spatial models become actionable. The implementation of AIS provides an example of including additional information as resources to use during navigation. Important for these changes is that forms of movement are particularly important for making observations at all. Temporal data provides a means to synchronize this information, allowing to gather, or "scope" information that is outside of the reach of near-immediate visual or auditive cues. Instead of simulating a simultaneity, charts and AIS information on plotters tends to be updated partially, so that for example changes in wind force, direction, or the direction of vessels around oneself may have broad temporal discrepancies.

While the High North Atlantic is only one area within which navigational images are deployed, it is of particular importance especially for the history and future development of AIS. Navigation, as a way of relating to the winds, relies particularly on kairos, and the use of different modes of navigation, in order to find one's way.

Notes

1 Technical details concerning AIS can be found in the International Maritime Organization's *SOLAS regulation V/19* as well as in the International Telecommunications Union's *Recommendation ITU-R M.1371.*
2 Information on NorSat-satellites can be found e.g. on the website of the *European Space Agency*, see e.g. https://www.esa.int/Applications/Connectivity_and_Secure_Communications/Satellites_paint_a_detailed_picture_of_maritime_activity
3 Some scholars expand the notion of map to include mental images (e.g. Edney, 2017), or take up the notion of mappings in order to pay attention to the processuality of creating and using maps (Kitchin and Dodge, 2007). While I am sympathetic to these suggestions, I consider navigation to be an activity which is not constrained to maps nor mappings.
4 For a discussion of the distributed cognition needed for navigation onboard, see Hutchins, 1995.
5 The terms operative and operational images have been used to describe machinic ways of processing and analysing images, at times in ways in which the images themselves are never visualized for humans to seen (cf. Farocki, 2004; Paglen, 2019; Feiersinger, Friedrich, and Queisner, 2018). Navigational images may be operative, but do not need to be so. They are images that are used to orient oneself towards the future, which is in itself understood to be volatile and changing.
6 For a broader discussion of the importance of measuring time in order to know space, see Galison (2003) and Siegert (2015).
7 Denny (2012) explains a range of navigational techniques, including the functioning of Radar, VHF Omni Radio, Loran, and so on in an accessible way.
8 For a brief overview, see Cutlip (2017).
9 Cf. Helmreich (2019, p. 742) for the role of tragic floods for a perceived need to develop a wave height-measuring buoy.
10 See Recommendation ITU-R M.1371 and IMO: SOLAS regulation V/19—Carriage requirements for shipborne navigational systems and equipment.
11 See NASA (2014).
12 Herman Korsström published in 1911 a textbook in navigation used for education of navigational techniques, *Lärobok i navigation*, which is fittingly divided into two parts: The terrestrial and the astronomical. For research on the history of navigation before electronic computing, see Mörzer Bruyns (2009).
13 The Royal Museums Greenwich have useful resources on the history of the Royal observatory as well as the history of specific navigational techniques, see e.g. https://www.rmg.co.uk/stories/topics/harrisons-clocks-longitude-problem
14 Only in 1967 was it internationally agreed to change the length of a second to be based on properties of the caesium atom, after which time is only loosely bound to celestial bodies, focusing rather on an electronic device counting vibrations and converting the count rate into a time on a clock face. For an example, see https://www.rmg.co.uk/collections/objects/rmgc-object-79654

References

Bauer, K. (2009). On the politics and the possibilities of participatory mapping and GIS: Using spatial technologies to study common property and land use change among pastoralists in Central Tibet. *Cultural Geographies*, 16 (2), 229–252. doi:10.1177/1474474008101518.

Belting, H. (2007) Blickwechsel mit Bilder. Die Bildfrage als Körperfrage. In H. Belting (Ed.) *Bildfragen. Die Bildwissenschaft im Aufbruch* (pp. 48–75). Fink.

Belting, H. (2011) *An Anthropology of Images: Picture, Medium, Body.* Princeton University Press.

Caquard, S. & Cartwright, W. (2014). Narrative cartography: From mapping stories to the narrative of maps and mapping. *The Cartographic Journal*, 51 (2), 101–106. doi:10.1179/0008704114Z.000000000130.

Denny, M. (2012). *The Science of Navigation. From Dead Reckoning to GPS.* Johns Hopkins University Press.

Edney, M. H. (2017). Map history: discourse and process. In A.J. Kent & P. Vujakovic (Eds.). *The Routledge Handbook of Mapping and Cartography* (pp. 68–79). Routledge.

Elden, S. (2013). Secure the volume: Vertical geopolitics and the depth of power. *Political Geography*, 34 (May), 35–51.

Farocki, H. (2004). Phantom images. *Public*, 29, 12–22.

Feiersinger, L., Friedrich, K., & Queisner, M. (Eds.) (2018). *Image–action–space: Situating the Screen in Visual Practice.* Walter de Gruyter.

Galison, P. (2003). *Einstein's Clock's, Poincaré's Maps.* W.W. Norton.

Grasseni, C. & Gieser, T. (2019). Introduction: Skilled mediations. *Social Anthropology*, 27 (1), 6–16.

Helmreich, S. (2019). Reading a wave buoy. *Science, Technology, & Human Values*, 44 (5), 737–761. doi:10.1177/0162243919856095.

Hutchins, E. (1995). *Cognition in the Wild.* MIT Press.

Ingold, T. (2002). Stop, look and listen! Vision, hearing and human movement. In T. Ingold. *The Perception of the Environment: Essays on Livelihood, Dwelling and Skill* (pp. 243–287). Routledge.

Ingold, T. (2016). *Lines: A Brief History.* Routledge.

Kitchin, R. & Dodge, M. (2007). Rethinking maps. *Progress in Human Geography*, 31 (3), 331–344. doi:10.1177/0309132507077082.

Koch, T. (2004). The map as intent: Variations on the theme of John Snow. *Cartographica: The International Journal for Geographic Information and Geovisualization*, 39 (4), 1–14. doi:10.3138/B123-8124-4390-5792.

Knorr Cetina, K. (2009). The synthetic situation: Interactionism for a global world. *Symbolic Interaction*, 32 (1), 61–87. doi:10.1525/si.2009.32.1.61.

Knorr Cetina, K. (2014). Scopic media and global coordination: The mediatization of face-to-face encounters. In K. Lundby. *Mediatization of Communication* (pp. 39–62), De Gruyter. doi:10.1515/9783110272215.39.

Korsström, H. (1911). *Lärobok i navigation.* Stadsrådet.

Lynch, K. (1960). *The Image of the City.* Technology Press & Harvard University Press.

Mörzer Bruyns, W. F. J. (2009). Research in the history of navigation: Its role in maritime history. *International Journal of Maritime History*, 21 (2), 261–286. doi:10.1177/084387140902100212.

November, V., Camacho-Hübner, E., & Latour, B. (2010). Entering a risky territory: Space in the age of digital navigation. *Environment and Planning D: Society and Space*, 28 (4), 581–599.

Paglen, T. (2019). Invisible images: Your pictures are looking at you. *Architectural Design*, 89 (1), 22–27.

Parikka, J. (2023). *Operational Images: From the Visual to the Invisual.* University of Minnesota Press.

Peters, J. D. (1997). Seeing bifocally: Media, place, culture. In A. Gupta & J. Ferguson. *Culture, Power, Place: Explorations in Critical Anthropology* (pp. 75–92). Duke University Press.

Peters, J.D. (2020). *The Marvelous Clouds: Toward a Philosophy of Elemental Media.* University of Chicago Press.

Pezzani, L., & Heller, C. (2019). AIS politics: The contested use of vessel tracking at the EU's maritime frontier. *Science, Technology, & Human Values*, 44 (5), 881–899. doi:10.1177/0162243919852672.

Rose-Redwood, R. (2015). Introduction: The limits to deconstructing the map. *Cartographica: The International Journal for Geographic Information and Geovisualization*, 50 (1), 1–8. doi:10.3138/carto.50.1.01.

Siegert, B. (2015). Longitude and simultaneity in philosophy, physics, and empires. *Configurations*, 23 (2), 145–163.

Siegert, B. (2023). *Kulturtechniken.* Rombach Wissenschaft.

Thielmann, T. (2010). Locative media and mediated localities: An introduction to media geography. *Journal of Media Geography* 5 (1).

Thielmann, T. (2018). Early digital images. A praxeology of the display. In L. Feiersinger et al. (Eds.) *Image–action–space: Situating the Screen in Visual Practice* (pp. 41–54). Walter de Gruyter.

Verhoeff, N. (2012). *Mobile Screens: The Visual Regime of Navigation.* Amsterdam University Press.

Vigh, H. 2009. Motion squared: A second look at the concept of social navigation. *Anthropological Theory*, 9 (4), 419–438. doi:10.1177/1463499609356044.

Virilio, P. (2007). *The Original Accident.* Polity.

Reports

International Maritime Organization. *SOLAS regulation V/19 - Carriage requirements for shipborne navigational systems and equipment.*

International Telecommunications Union. *Recommendation ITU-R M.1371. Technical characteristics for a universal shipborne automatic identification system using time division multiple access in the VHF maritime mobile band.*

Websites

Cutlip, K. 2017AIS for safety and tracking: A brief history. *Global Fishing Watch*, https://globalfishingwatch.org/data/ais-brief-history/ (last accessed March 31, 2024).

European Space Agency. Satellites paint a detailed picture of maritime activity. https://www.esa.int/Applications/Connectivity_and_Secure_Communications/Satellites_paint_a_detailed_picture_of_maritime_activity (last accessed March 31, 2024).

NASA2014. ISS benefits for humanity: Found at sea. Short documentary. In *Internet Archive*, https://archive.org/details/ISS-Benefits-For-Humanity_Found-at-Sea (last accessed March 31, 2024).

Royal Museum Greenwich. Atomic clock. https://www.rmg.co.uk/collections/objects/rmgc-object-79654 (last accessed March 31, 2024).

Royal Museums Greenwich. Longitude found: The story of Harrison's clocks. https://www.rmg.co.uk/stories/topics/harrisons-clocks-longitude-problem (last accessed March 31, 2024).

8 Sea Ice Extent and the Sea Ice Index

Freezing Arctic Sea Ice through Measurement and Mediation[1]

Laura Seddon and Philip Steinberg

Introduction

As is illustrated throughout this volume, the High North Atlantic and its adjacent seas are increasingly understood via mediations: images and imaginaries that, by conveying information, construct realities. The images shown in Figure 8.1, taken from the U.S. National Snow and Ice Data Center (NSIDC) Sea Ice Index, exemplify this trend. The Sea Ice Index often serves as the "benchmark dataset" for sea-ice extent and is one of the most visible and compelling indicators of global environmental change, reduced to a single variable (Meier and Stewart, 2019; WMO, 2022). This series of images has had an outsized impact on the discursive framing of northern oceans, and indeed the entire world, constructing it as, paradoxically, both on the cusp of environmental catastrophe and ripe with economic opportunity (Christensen and Nilsson, 2019).

In this chapter, we discuss how the Sea Ice Index, and the principle of sea-ice extent that underpins it, is a communicative tool that is neither inevitable nor wholly objective but dependent on the situated arrangements and interactions between sea ice, technologies, and people that prevail in specific cultural contexts and at particular moments in time. Our critique thus resonates with, and builds upon, works of others who have stressed how the ways that we gather, communicate, and mediate scientific knowledge are socially and culturally contingent, reflecting and reinforcing particular values, perspectives, and interests while marginalizing others (e.g., Latour and Woolgar, 1979; Jasanoff, 2004a; Harding, 2015), as well as the critique made more specifically about the social mediation of climate science (e.g., Jasanoff, 2004b; Findlater et al., 2021; Lahsen & Turnhout, 2021).

However, our critique also turns expressly to an investigation of the foundational technology of the Sea Ice Index: the satellite. Often termed the "eyes in the sky" (Wormbs, 2013), a satellite's distanced position provides an exceptional perspective on the world (Cosgrove, 1994; Parks, 2005) and this exceptionality has played a powerful role in the notion that satellites are uniquely able to objectively observe "everything from nowhere" (Haraway,

DOI: 10.4324/9781003478805-8

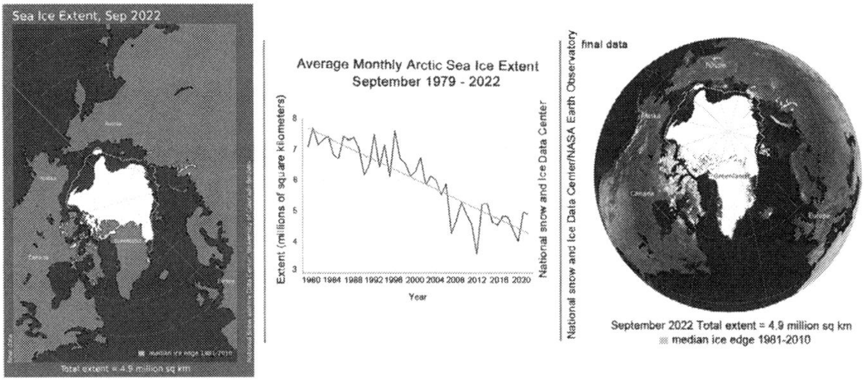

Figure 8.1 Maps and graphs of sea-ice extent taken from the National Snow and Ice Data Center (NSIDC) Sea Ice Index
Source: https://nsidc.org/data/seaice_index

1988: 581). Consequently, satellites, together with digital mapping, have become a dominant way of "seeing" and documenting changes in the global environment, elevating some of its most "visible" elements—including the presence or absence of sea ice—to iconic status. The Sea Ice Index, in concert with the satellites that enable its production and the images that communicate its meaning, mediates northern waters, constructing them as a space of environmental change and economic potential.

The chapter begins by introducing the role of sea ice in the climate system and its particular significance for the High North Atlantic as well as the adjacent Arctic Ocean. Then, the chapter critically examines the technologies of mediation involved in deriving estimates of sea-ice extent using satellites. These critical examinations draw primarily from the analysis of publicly available sea-ice extent data products and their surrounding documentation, which includes visual and statistical elements (e.g., sea-ice maps and graphs), supplementary documentation (e.g., product user guides and technical reports), and other supporting textual materials and statistical data (e.g., institutional webpages, published literature, and government policy reports).[2] Collectively, these documents are both a part of and contain essential information about the mediating technologies underpinning sea-ice extent as a measure of environmental change. The analytical focus involves paying close attention to the data retrieval methods and processing techniques, assumptions, conventions, and uncertainties that shape how sea ice is interpreted, defined, and ordered in the production and communication of scientific knowledge.

Through these examinations, the chapter problematizes the dominant understanding of sea-ice extent as a seemingly straightforward indicator by highlighting the partial perspective it gives rise to. Rather, the Sea Ice Index is analyzed as reflecting a particular way of encountering and knowing northern oceans that is dependent on the situated arrangements and interactions between sea ice, scientific and technological practices, and people that prevail in specific socio-cultural and historical contexts. In questioning the authority of satellite observations and the power they have in shaping knowledge of and responses to environmental change, this chapter draws attention to the potential silences, exclusions, and inequalities that are reflected and reinforced by the dominant knowledge infrastructures of contemporary sea-ice science. We argue that these dominant approaches risk silencing the diverse socio-political, cultural, and ecological functions that sea-ice serves as well as marginalizing other ways of knowing and encountering sea-ice environments. The chapter concludes by discussing the need for more reflexive and pluralistic understandings of complex environments, which are essential to the development of more democratic and sustainable responses to environmental changes.

Sea-Ice and the Climate

Over recent decades, Arctic sea-ice cover has been declining at an unprecedented rate. The consensus of leading scientists is now stark and terminal: as Arctic temperatures continue to rise at nearly four times the global average rate, the region will experience ice-free summers within the coming decades (IPCC, 2021).[3] This environmental transformation has been described as *inevitable* and essentially *irreversible* and its potential climatic, ecological, and societal consequences are significant and global in scale (ICCI, 2022).

The imminent disappearance of perennial (multi-year) sea ice—a distinct feature of the Arctic that has persisted for millennia—serves as a clear reminder of the central role that humans have played in influencing Earth systems and material elements. Sea-ice decline is both a symptom and an amplifier of climate change, and its loss has significant implications. At a local and regional level, retreating sea ice results in ecosystem changes, such as altered food webs and migration patterns (Hauser et al., 2017), and the opening up of new shipping routes and increased potential for oil and gas exploitation (Petrick et al., 2017; Lasserre and Faury, 2019), all of which would have significant implications for communities whose culture, livelihoods, and well-being are deeply connected to the presence of sea ice (Durkalec et al., 2015). Declines in sea ice also amplify climatic changes through modulating atmospheric and oceanic exchanges, Earth's surface albedo, and ocean circulation (e.g., Flanner et al., 2011).

Declining sea ice has had a particularly profound impact on the northern oceans, a term that we use here to bridge the linked spheres of the Arctic Ocean (where sea ice has historically been prevalent) and the High North Atlantic (which has been free of sea ice). Atmospheric and oceanic exchanges between the Arctic Ocean and the High North Atlantic are critically important for the climate of the Arctic region and the state of sea-ice cover. Growing evidence indicates that increasing air and sea-surface temperatures in subarctic regions (including the High North Atlantic) are altering the climatic features and mechanisms of the Arctic Ocean (Ingvaldsen et al., 2021; Polyakov et al., 2023). Warmer, saltier subarctic water is extending northwards into the Arctic Ocean, weakening oceanic stratification and accelerating sea-ice loss. This process, called "atlantification," is causing areas of the Arctic Ocean to resemble more closely the Atlantic, in terms of physical, environmental, and ecological characteristics.

Scientists and research institutions worldwide are monitoring this loss of sea ice, using observations obtained through various *in situ* and remote sensing methods and transforming them into maps, statistics, graphs, and indices that are ubiquitous in popular and scientific media. In particular, measurements of the total areal coverage (*extent*) of sea ice have become a dominant focus in climate change discourse (e.g., WMO, 2022). Defined as the total surface area covered by an ice concentration above a quantitative threshold (usually 15%), sea-ice extent is a widely accepted scientific parameter used to characterize the state of sea-ice cover and monitor its long-term and large-scale variability. Knowing how much ocean is covered by sea ice at any one time is considered essential for understanding the role that sea ice plays in the global climate system, assessing the potential implications of continued melting, and providing the empirical basis for informing appropriate action, decision-making, and policy responses.[4]

Localized interest in sea ice has existed for millennia, particularly in northern oceans. As a key substance for facilitating Indigenous mobilities, hindering outsiders' mobilities, and nourishing the algae that forms the foundation of the marine food chain, understanding and accounting for the presence and absence of sea ice has long been necessary for cultural survival. However, the use of a single figure to indicate sea-ice extent to define and monitor the regional, hemispheric, and even global presence of sea ice has a much shorter history. Technological developments after World War Two, such as advances in computational technologies and monitoring systems and new methods of scientific inquiry, accelerated a shift toward more collaborative and synchronous scientific research into the climate, which was increasingly characterized by quantified, modeled, and deductive analytical methods (Doel, 2003). The strategic need for predictive knowledge during the Cold War led to the establishment of numerical weather prediction, and later, global climate models that rendered complex climatic phenomena calculable to simulate past, present, and future conditions (Edwards, 2010).

Although theoretical and model-based literature on the climatic importance of Arctic sea ice swelled during the 1960s and 1970s (e.g., Parkinson and Kellogg, 1979), its robustness was limited by a discontinuous catalogue of empirical observations. Historical (pre-satellite) observations of sea ice were obtained by ground-, ship-, and later, air- and submarine-based measurements (Seddon, 2022), but the North's extreme climate and dark winters, as well as the transient and expansive nature of sea ice, presented significant barriers to obtaining empirical observations on a sufficient frequency and scale to model sea-ice variability on climatic timescales (Makyut, 1986).

Polar observation underwent a major transformation with the satellite era, which, although beginning in the 1950s, became particularly significant for sea-ice observation with the launch of NASA's suite of Nimbus satellites during the 1970s. Characterized by a polar-orbiting, Earth-oriented configuration, these satellites enabled almost complete, global coverage of the Earth's surface. Advances in electromagnetic imaging technologies also enabled the satellites to be equipped with passive microwave radiometers (PMRs), which detect and measure the microwave radiation naturally emitted by the Earth's surface, from which surface characteristics can be estimated (Tedesco, 2015). Observations of passive microwave radiation are particularly advantageous over other observational methods because clouds emit and scatter very small amounts of microwave radiation and the microwave radiation emitted and scattered by sea ice is not dependent on solar illumination (Eppler et al., 1992). This means that PMRs can retrieve surface observations irrespective of the amount of cloud cover or daylight, permitting a record of sea-ice observations far more complete and consistent than available by other observational methods (Meier and Markus, 2015).

The introduction of PMRs brought about a fundamentally different approach to sea ice, practically and epistemically. Notably, the observations acquired using PMRs are not direct measurements of sea ice. PMRs measure the microwave radiation naturally emitted by the surface, from which surface types can be distinguished using electromagnetic theory and the application of classification algorithms (Meier and Markus, 2015). With the introduction of PMRs, monitoring sea-ice characteristics no longer depended solely on scientific expeditions for *in situ* ground-, air-, or sea-based measurements. Instead, it relied upon the treatment of sea ice as an abstract and calculable unit, the characteristics of which could be estimated through empirically-derived assumptions and digital processing. Often this mediation of the sea ice's presence occurred in offices far removed from the ice being "measured."

Except for a small observational gap at the North Pole, PMR satellite-sensors provided near-complete coverage of northern oceans every 24–48 hours.[5] No such consistent, spatially extensive data had been available from other means of sea-ice observation—a statement that remains broadly true today. To meet the needs of "user" communities seeking to integrate PMR-derived sea-ice data (e.g., researchers studying ocean-atmosphere processes), NASA developed systems and techniques for acquiring, processing, validating, and translating satellite data into scientifically useful sea-ice parameters, using standard protocols, thresholds, and definitions (Cavalieri, Gloersen, and Campbell, 1984). These parameters included sea-ice concentration and extent and the position of the sea-ice edge, datasets of which were archived and distributed by the NSIDC, and these forms of data translation became a principal tool through which scientists interpreted and reported satellite observations of sea ice, including through their visualization (e.g., Cavallieri and Zwally, 1985; see also Figure 8.1). Although simplistic in their design, maps and graphs of sea ice quantities provided a revolutionary perspective on

Arctic sea ice: their seemingly all-encompassing, bird's-eye view of the Earth, facilitated by satellites, stood in stark contrast to the temporally and spatially sparse observations collected during preceding centuries.

Over time, continuous, daily PMR observations of Arctic sea ice have documented significant and long-term downward trends in sea-ice extent, shifts in the seasonality of sea ice, and changes in sea-ice composition (e.g., Cavalieri and Parkinson, 2012). These observations have helped facilitate the shift toward understanding sea ice as a *bellwether* of global climate change that demands urgent attention and action (Christensen and Nilsson, 2019). Likened to a paradigm shift in science (Jasanoff, 2004b), the global view satellites offer has transformed our knowledge of northern oceans, including the challenges and opportunities related to future change and how these might be addressed. In this way, satellite observations have come to be seen not simply as data but as part of a global solution to the problem of sea-ice decline (GCOS, 2018).

Mediating Icy Worlds

Today, PMR observations are used by scientists worldwide to develop consistent, comprehensive, and homogenized sea-ice datasets and data products that inform the scientific community, policymakers, industry, and the public. Sea-ice extent is now regarded by the Global Climate Observation System as an "essential climate variable," one of the physical, chemical, or biological variables that indicates climactic variability and for which there are standardized requirements regarding data quality, including observational frequency, accuracy, and resolution (Seddon, 2022).

Observations of sea-ice extent cover a period of over 40 years, supplying the datasets necessary for scientists to analyze changes in sea-ice cover on regional, seasonal, and inter-annual timescales (e.g., Cavalieri and Parkinson, 2012). Supporting these expanding datasets, developments in infrastructure and computation have permitted satellite observations to be processed, analyzed, and disseminated in more efficient and innovative ways: visualizations of sea-ice extent appear detailed, vibrant, realistic, and animated. A notable example of this is the NSIDC's representation of sea-ice extent using NASA's *Blue Marble view of the Earth* (Figure 8.1), which is used to communicate sea-ice conditions in a way that is accessible and understandable to a variety of audiences (Fetterer and Knowles, 2004). Graphic, numerical, and textual depictions of sea-ice extent using NSIDC criteria and modes of representation are commonplace in scientific and public media, reducing complex and indeterminate processes of sea-ice decline to singular lines, whether in the Norwegian media's reduction of the Marginal Ice Zone to a singular "ice edge" (derived from NSIDC data) (Steinberg, Kristoffersen, and Shake, 2020) or in the depiction of a "maximum winter extent" line (again derived from NSIDC data) that is now standard on National Geographic maps.[6]

Sensing, Gridding, and Classifying Sea-Ice Extent

Notwithstanding the ubiquity of PMR-derived calculations and visualizations as mediations of sea-ice (and hence environmental) knowledge, in their raw state PMR observations have little value for monitoring or measuring sea-ice conditions. As was noted above, satellite PMRs do not directly measure the geophysical properties of sea ice. Rather they gather data—derived from the different electromagnetic properties of different surfaces—that can be used to calculate the *radiation signature* (or, *brightness temperature*) which, in turn, can be used to identify the dominant surface type (Eppler et al., 1992). In other words, the production of meaningful sea-ice extent data from PMR observations involves a complex and multifaceted transformation process involving several stages of selection, interpretation, conversion, and simplification (Tedesco, 2015). In the process, the spatial information sensed by the satellite becomes increasingly mediated.

To perform these mediations, scientists have developed a range of calculative algorithms, more than 30 of which are in use today (Pedersen et al., 2017). While these algorithms differ in their technicalities, approach, and prevalence, all essentially share a common assumption: a combination of microwave emissions from first-year ice, multi-year ice, and open water comprises the brightness temperature measured by the sensor (Meier and Stewart, 2019). Most sea-ice retrieval algorithms exploit this assumption to approximate the *sea-ice concentration* (fractional coverage) over a given area. Sea-ice concentration forms the basis of sea-ice extent calculations. In contrast to sea-ice concentration, which is communicated as a percent, sea-ice extent is a binary measure that classifies a region as either "ice-covered" or "ice-free," depending on whether the sea-ice concentration exceeds or falls short of a threshold definition, usually set at 15%. The 15% threshold became accepted practice following validation studies carried out during the 1980s, which compared PMR observations against higher-resolution aircraft-based observations (Cavalieri et al., 1991). These studies found that the location of the outer edges of sea-ice cover identified from the air corresponded best with the satellite-derived 15% sea-ice concentration. However, the initial choice of threshold definition was complicated by the disaggregated nature of the sea-ice edge.

Daily averaged brightness temperature data are transformed into calculations of sea-ice concentration and extent using *gridded data fields*. These fields organize the data into discrete cells corresponding to specific ocean surface areas. PMR observation data are typically mapped to grid cells with a gridded resolution of 25 km (Figure 8.2). Daily sea-ice extent values are calculated by summing the *total* area of all grid cells with sea-ice concentrations above a defined threshold. Sea-ice maps use single-block colors to classify grid cells into "ice-covered" or "open-water" areas, creating a static and linear ice-edge and homogenous divisions between surface types. These calculations and procedures form the basis of maps, graphs, and values that capture sea-ice extent on daily to decadal timescales.

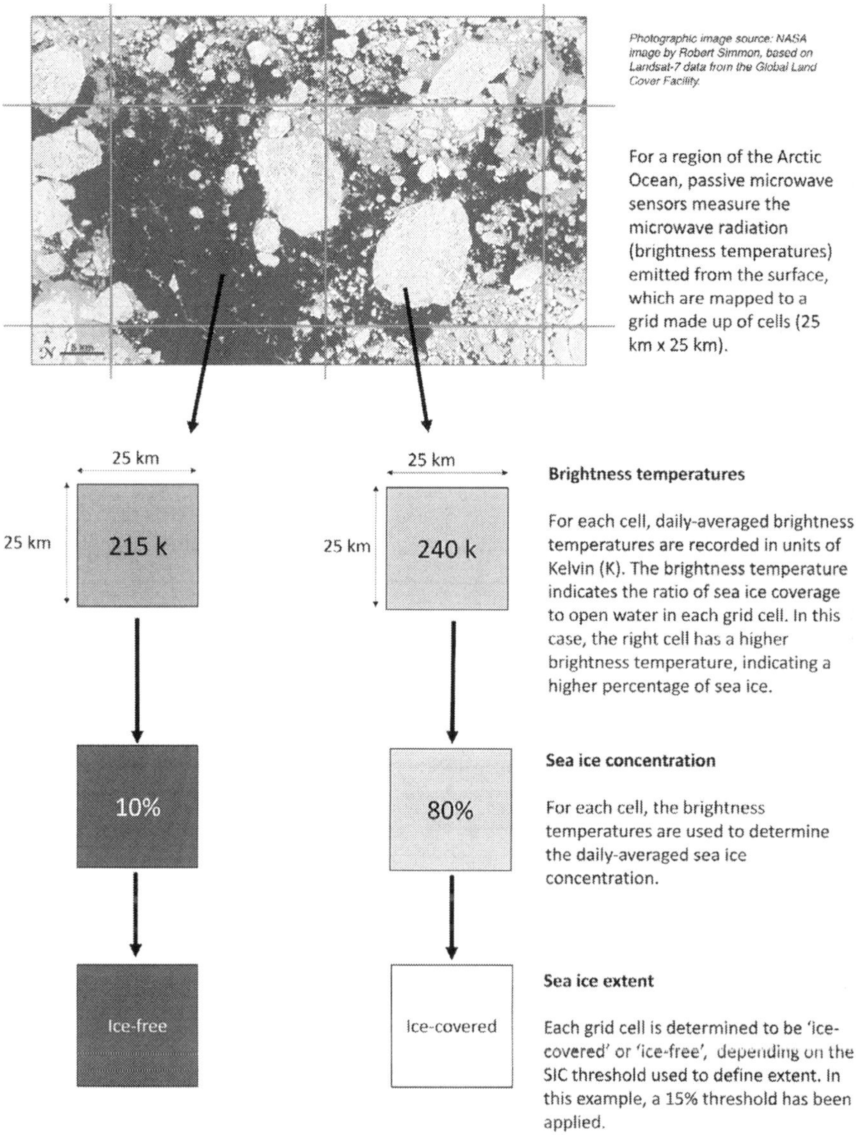

Figure 8.2 Schematic diagram showing the key processes and definitions involved in determining sea-ice extent measurements from passive microwave satellite observations. Note that the values and symbols used here are purely indicative and not precise measurements of sea-ice conditions.

Uncertainties and Noisy Signals

For over 30 years, sea-ice extent calculations derived from PMR sensors have dominated scientific understanding of sea-ice cover: they have been used to determine the absolute quantities of sea ice, where the edge is located, and what types of sea ice are present (e.g., first-year or multi-year). They have also been crucial to informing scientific understanding of potential futures for the northern oceans; for example, speculations about when the Arctic Ocean will experience its first ice-free summer are dependent on sea-ice extent calculations (Notz and Stroeve, 2018). However, various limitations characterize satellite PMR technologies, introducing uncertainties and ambiguities in the measurement and parameterization of sea ice.

A significant limitation is the coarse spatial resolution of PMRs, defined by the sensor's *footprint*: the smallest object or area of the surface that a sensor can capture at any given moment. The footprint—a technical term unrelated to the "environmental footprints" more commonly referenced in climate policy—is inversely proportional to the microwave frequency (amount of energy) being sensed (Pedersen et al., 2017). Because the microwave energy emitted from the Earth's surface is characteristically low, PMRs require large footprints to capture sufficient energy for surface characterization (Meier and Markus, 2015). Here, it is important to clarify the difference between the previously mentioned *gridded resolution* and the *spatial resolution of the PMR*. The former represents the resolution to which satellite observations are aggregated during data processing—a necessary step for parameterizing and quantifying sea-ice characteristics. The actual spatial resolution of the observations is often significantly coarser, ranging up to several tens of kilometers. Because of this, when a satellite PMR "senses" the surface, it detects and measures the integrated radiation signatures (brightness temperatures) over a large surface area. These integrated signals can obscure, generalize, and simplify surface-level complexities and characteristics; smaller-scale features, such as leads, polynyas, and localized variability, often remain "unseen" (Meier and Markus, 2015).

Various atmospheric and oceanic features, such as sea-surface roughness and atmospheric water vapor, can further complicate the measured microwave signatures (Meier and Markus, 2015). Collectively, these features represent unwanted signals that attenuate, contaminate, and replicate the brightness temperatures of sea ice. This *noise* can make it difficult to accurately differentiate between open water and sea ice and undermine the transformation of radiation data into quantifiable sea-ice parameters. Additional correction methods can mitigate—but not altogether remove—the effects of some of these uncertainties and ambiguities on approximations of sea-ice cover (Cavalieri et al., 1999).

The uncertainties introduced by coarse spatial resolution and noisy signals are compounded by differences in data processing methodologies. Various institutions are involved in producing, representing, and communicating sea-ice extent using PMR observations.[7] Despite, in most cases, drawing from the same catalogue of PMR observations and adopting the same fundamental principles of satellite passive microwave remote sensing, the exact methodologies used to select, process, aggregate, and transform the satellite observations into ordered and meaningful knowledge about sea ice vary between institutions (Seddon, 2022). Some of the more notable differences include the choice of sea-ice retrieval algorithm, the frequency channels selected to obtain brightness temperatures, and the corrective procedures applied to reduce errors and uncertainties. A consequence of diverging methodologies is that, although the seasonal variability and long-term declining trends in sea ice align indisputably between different data producers, absolute sea-ice concentration and extent estimates can vary substantially: recent studies have demonstrated that sea-ice extent values vary by as much as 1 million km^2 (e.g., Comiso, Meier, & Gersten, 2017; Meier and Stewart, 2019); the mapped "ice edge" position at 15% sea-ice concentration can also vary by as much as 25 to 75 km (Meier and Stewart, 2019).

Many methodological differences reflect varying ways of dealing with the inherent uncertainties and limitations of satellite PMR observations. In other words, they reflect varying attempts to make the final outputs more representative—or "truth-like." However, scientists' attempts to derive more accurate estimates of sea-ice conditions are impeded by *the absence of "truthful" data*. In remote sensing, *"ground-truthing"* is an essential step for validating, verifying, and calibrating spaceborne observations against actual features at the surface. Ground-truth data on sea ice includes *in situ* measurements, surface-level conditions obtained from ships and on-ice expeditions, and other higher-resolution forms of remote sensing. However, such observations are spatially and temporally sparse due to the challenges of environmental observation in the northern oceans. Moreover, while these higher-resolution datasets may be considered closer approximations of reality, they are still just that: *approximations*. They, too, are characterized by various uncertainties and assumptions and involve several stages of data aggregation, simplification, and manipulation which interact with and mediate sea-ice conditions in varying ways. This means that there is no definitive way of assessing the absolute accuracy (or "truthfulness") of PMR observations of sea ice—at least not within the confines of contemporary science and technology. There is no single, universal account of sea-ice conditions against which sea-ice parameter estimates can be compared and no way of knowing exactly how much sea ice exists at any one time. This has facilitated the development of different methodologies used to process and represent PMR observations of sea ice. Using Harry Collins' words, it has enabled a certain degree of *interpretative flexibility* in producing and disseminating sea-ice knowledge (Collins, 1985).

Partialities of Sea-Ice Extent

As is the case with climate science more broadly, uncertainties or ambiguities in scientific findings challenge the translation of knowledge into policy action as they can prompt questions over the veracity of scientific claims and present multiple problems and potential solutions. Several intercomparison studies have been carried out to identify and quantify uncertainties in sea-ice extent calculations and reduce the causes of discrepancies (e.g., Ivanova et al., 2015; Comiso, Meier, and Gersten, 2017; Meier and Stewart, 2019). These ongoing efforts imply that the sea-ice science community is committed to achieving a *singular*, consensus-driven knowledge of Arctic sea-ice conditions that embodies the objectivist ideal suggested by the satellite: a *view from nowhere* (Nagel, 1986). From a critical perspective, however, the divergent methodologies and resulting outputs emphasize the multiplicity of potential interpretations and meanings that can emerge from satellite observations of sea ice.

To be clear, we do not oppose the use of sea-ice extent as a measure for assessing and interpreting climate change. In fact, the practices and technologies involved in deriving sea-ice parameters build upon several decades of scientific research and consensus, and scientists produce and communicate sea-ice extent estimates as part of a global and systematic effort to understand and monitor environmental change in northern oceans. In this respect, the importance of sea-ice extent for capturing long-term and large-scale change is hard to refute. Nonetheless, the Science and Technology Studies literature referenced at the beginning of this chapter suggests that we need to be skeptical of how the Sea Ice Index is deployed. Our critique is not *just* that the attempt to define a singular, supposedly objective, universal measure of sea-ice extent and endow it with transcendent qualities for mediating climate processes reproduces the power of specific scientific institutions. Nor is it *just* that the singularity of the Sea Ice Index masks uncertainty and the arbitrary, contested assumptions that underpin the assignation of a single binary variable to a complex environment. For us, the problem lies, most profoundly in the fact that the Sea Ice Index mediates how we know this environment, and, in the process, it legitimates and constrains the very nature of knowledge and the universe of potential responses. As this partial, selective, and technical mediation of sea ice becomes *the* dominant way of knowing the changing environment in the northern oceans, and as alternate knowledges derived from experience and practice decline, or lose legitimacy, or are subsumed into scientific frameworks, the Sea Ice Index—whether presented as a percentage, a measurement of square kilometers, a falling data point on a graph, or a mapped line—increasingly operates as a "more-than-scientific reference" (Gabrys and Yusoff, 2012, p. 6).

Attempts to derive a (seemingly) straightforward indicator of geophysical change come at the expense of the dynamism, material complexity, and diverse functionality and meanings of sea-ice environments. Sea-ice extent defines, constructs, and stabilizes (temporally and spatially) boundaries between the open ocean and ice-covered waters that do not exist in reality, at least not in a distinct, linear, and uncontested form (Steinberg, Kristoffersen, and Shake, 2020). A static boundary between ice and water does not characterize the "ice edge"; it is a zone of transition that can extend over several tens of kilometers and is in constant motion due to wind and ocean currents.

In this zone, sea ice exists in many different material forms, and diverse geophysical, biological, chemical, and social interactions and interchanges take place. Fixing boundaries between "ice-free" and "ice-covered" areas is further complicated by the various and often conflicting functions sea ice serves. As Steinberg and Kristoffersen write: "the same area of sea ice may be a hazard to a ship, a highway for a dog sled or snowmobile and a place of refuge for a polar bear" (2017, p. 629). Across these different contexts, sea ice may simultaneously represent a barrier to movement, a facilitator of movement, or a habitat below, above, or within the ice. Thus, what constitutes "ice-covered" varies depending on the circumstances of the encounter. However, this multiplicity of meaning can never be captured by an index that aggregates across a single, binary value that itself is derived from contested algorithms and arbitrary assumptions.

There are alternative ways of knowing sea ice, and these persist in the practices of people who engage ice through practiced encounters, whether as Indigenous hunters or as whalers or as scientists working through other methodologies in other scientific traditions. As just one counter-example, the *Siku-Inuit-Hila* (Sea Ice-People-Weather) project that brought together social and physical scientists and members of Indigenous communities in Utquiaviq (Barrow), Alaska; Kangiqtugaapik (Clyde River), Nunavut; and Qaanaaq, Greenland was notable not just for sharing ideas, knowledge, skills, and experiences relating to sea ice in order to develop a greater understanding of the dynamics of the relationships between sea ice and humans in northern oceans. It also was notable because of the form of its output—*The Meaning of Ice* (Gearheard et al., 2013)—a rich collection of descriptions, photographs, artwork, illustrations, maps, glossaries, and stories that were generated through the scientist-Indigenous collaboration. The collection provides comprehensive insights into the dimensions of life in northern oceans, the diverse role that sea ice plays in shaping livelihoods, and the various meanings that people attach to sea ice. As such, it challenges received understandings of *knowledge*. Notably, the knowledge presented in this collection is often at odds with simplistic, standardized, and impersonal scientific representations exemplified by the Sea Ice Index. Instead, it is diverse, subjective, evocative, and personal.

To summarize, the emergence of the Sea Ice Index as a dominant means for communicating changing conditions within linked marine environments, in the Arctic Ocean, the High North Atlantic, and beyond, does more than reflect the hegemony of Western science. It also reflects a tendency in climate science to translate and condense complex scientific findings into "quick-look" and easily interpreted representational forms (Schipper, Dubash, and Mulugetta, 2021). While this may be a necessary step for achieving a quantifiable parameter, delineating northern oceans into distinct and homogenous ice-covered and open-water areas perpetuates a highly mediated, abstract, and inherently partial view. Echoing other climate science critiques (e.g., Jasanoff, 2004b; Findlater et al., 2021; Lahsen and Turnhout, 2021), the privilege attached to quantitative, technical, and geophysical aspects subverts many of the broader socio-political, ecological, and cultural challenges and interests associated with sea-ice environments. At the same time, the scale of sea-ice extent measurements reinforces the framing of sea-ice decline as a global phenomenon requiring political action on a global scale. Even while helping to foster international collaboration and responsibility, this global-scale framing comes at the expense of more local discourses and lived experiences, which are often far more nuanced and complex (Brüggemann and Rödder, 2020; Eriksen, 2020). These subversions have direct implications for the types of phenomena made visible, knowable, and usable for different socio-political interests and agendas and can lead to the development of unsustainable and undemocratic policy responses to climate change (Findlater et al., 2021; Lahsen and Turnhout, 2021).

Concluding Discussion

The ways in which knowledge of Arctic sea ice is produced and disseminated have consequences for how we perceive, understand, and govern northern oceans. Satellite observations of sea-ice extent are produced and disseminated as part of a global effort to understand and communicate environmental change and mitigate against the effects of current and future climate change. However, as this chapter has argued, sea-ice extent is not simply a representation of surface conditions nor is it a straightforward indicator of climate change. Rather, satellite-derived sea-ice measurements serve a mediating function, reifying categories and thresholds that appear objective and natural but that, in fact, are established according to situated entanglements of environmental conditions, technological instruments, institutional contexts, and scientific and cultural interests. Consequently, satellite-derived observations of sea-ice extent and their associated representations mediate and condition sea-ice environments in ways that are inevitably partial, simplified, and contestable.

Therefore, there is a need for critical perspectives regarding the *situatedness* of sea-ice extent and the role it plays in articulating partial ways of knowing, and mediating, sea-ice environments (Seddon, 2022). To recognize that scientific knowledge of sea ice always comes from *somewhere* and is shaped and mediated by the contexts of its production is to recognize that knowledge is not inevitable. Although such critical perspectives imply that indicators like sea-ice extent can never be wholly objective and neutral, it does not mean that better and more equitable forms of knowledge cannot be realized.

This chapter contributes to the growing calls for climate change research that draws upon more interdisciplinary, pluralistic, and epistemologically diverse forms of knowledge (e.g., Schipper, Dubash, and Mulugetta, 2021). With the climatic, ecological, and societal implications of sea-ice decline already being felt globally, there is a need for technologies of mediation that better account for the complex materiality and dynamism of sea-ice environments, "un-freezing" the sea-ice science that has taken on such a crucial role in discourses surrounding global climate change as well as its impact in northern oceans.

Notes

1 This chapter is based on research funded by the Leverhulme Trust (Grant DS-2017–070 — Interdisciplinary Understanding for a Changing Arctic).
2 This chapter's analysis of these documents is derived from Seddon (2022).
3 Note that "ice-free" does not actually mean zero ice in absolute terms: "ice-free" is defined when total sea-ice extent is under 1 million km^2.
4 As is discussed at the end of this chapter, more localised sea ice monitoring is also used for routing the movement of everything from commercial shipping to Indigenous hunters, and this also frequently uses remotely sensed data.
5 Early sensors (1978–1987) operated on a bi-diurnal basis; since 1987, sensors have operated on a diurnal basis.
6 National Geographic World Maps use NSIDC SII sea-ice extent data to mark the position of the "ice edge": https://www.natgeomaps.com/wall-maps.
7 Examples include National Snow and Ice Data Center; Ocean and Sea Ice Satellite Application Facility; Japan Aerospace Exploration Agency; Institute of Environmental Physics, University of Bremen.

References

Brüggemann, M., and Rödder, S. (2020). We are climate change: Climate debates between transnational and local discourses. In M. Brüggemann and S. Rödder (ed). *Global Warming in Local Discourses: How Communities around the World Make Sense of Climate Change* (pp. 1–30). Cambridge, UK: Open Book Publishers.

Cavalieri, D.J., and Zwally, H.J. (1985). Satellite observations of sea ice. *Advances in Space Research*, 5(6), 247–255.

Cavalieri, D.J., Gloersen, P., and Campbell, W.J. (1984). Determination of sea ice parameters with the Nimbus 7 SMMR. *Journal of Geophysical Research: Atmospheres*, 89(D4), 5355–5369.

Cavalieri, D.J., Crawford, J.P., Drinkwater, M.R., Eppler, D.T., Farmer, L.D., Jentz, R.R., and Wackerman, C.C. (1991). Aircraft active and passive microwave validation of sea ice concentration from the Defense Meteorological Satellite Program Special Sensor Microwave Imager. *Journal of Geophysical Research: Oceans*, 96(C12), 21989–22008.

Cavalieri, D.J., Parkinson, C.L., Gloersen, P., Comiso, J.C., and Zwally, H.J. (1999). Deriving long-term time series of sea ice cover from satellite passive-microwave multisensor data sets. *Journal of Geophysical Research: Oceans*, 104(C7), 15803–15814.

Cavalieri, D.J., and Parkinson, C.L. (2012). Arctic sea ice variability and trends, 1979–2010. *The Cryosphere*, 6(4), 881.

Christensen, M. and Nilsson, A. E. (2019). *Arctic Geopolitics, Media and Power*. 1st edn. Taylor & Francis.

Collins, H.M. (1985), *Changing order: Replication and induction in scientific practice*. SAGE.

Comiso, J.C., Meier, W.N., and Gersten, R. (2017). Variability and trends in the Arctic Sea-ice cover: Results from different techniques. *Journal of Geophysical Research: Oceans*, 122(8), 6883–6900.

Cosgrove, D. (1994). Contested global visions: one-world, whole-earth, and the Apollo space photographs. *Annals of the Association of American Geographers*, 84(2), 270–294.

Doel, R. (2003). Constituting the postwar earth sciences: The military's influence on the environmental sciences in the USA after 1945. *Social Studies of Science*, 33(5), 635–666.

Durkalec, A., Furgal, C., Skinner, M.W., and Sheldon, T. (2015). Climate change influences on environment as a determinant of indigenous health: Relationships to place, sea ice, and health in an Inuit community. *Social Science & Medicine*, 136, 17–26.

Edwards, P. (2010). *A vast machine: Computer models, climate data, and the politics of global warming*. MIT Press.

Eppler, D.T., Farmer, L.D., Lohanick, A.W., Anderson, M.R., Cavalieri, D.J., Comiso, J.*et al.* (1992). Passive microwave signatures of sea ice. In F.D. Carsey (ed.). *Microwave remote sensing of sea ice* (pp. 47–68).American Geophysical Union.

Eriksen, F.C. (2020). The Case of "Costa del Nuuk": Greenlanders make sense of global climate change. In M. Brüggemann and S. Rödder (Eds.) *Global warming in local discourses: How communities around the world make sense of climate change* (pp. 31–76).Open Book Publishers.

Fetterer, F., and Knowles, K. (2004). Sea ice index monitors polar ice extent, *Eos Trans. AGU*, 85(16), 163–163. doi:10.1029/2004EO160007

Findlater, K., Webber, S., Kandlikar, M., & Donner, S. (2021). Climate services promise better decisions but mainly focus on better data. *Nature Climate Change*, 11(9), 731–737.

Flanner, M.G., Shell, K.M., Barlage, M., Perovich, D.K., and Tschudi, M.A. (2011). Radiative forcing and albedo feedback from the Northern Hemisphere cryosphere between 1979 and 2008. *Nature Geoscience*, 4(3), 151–155.

Gabrys, J., and Yusoff, K. (2012). Arts, sciences and climate change: Practices and politics at the threshold. *Science as Culture*, 21(1), 1–24.

GCOS. (2018). *Systematic observations and the Paris agreement: Report of the Task Team on the Paris Agreement*. GCOS-222, WMO. https://library.wmo.int/index.php?lvl=notice_display&id=20689#.YPk4Xi1Q1hE.

Gearheard, S., Holm, L.K., Huntington, H., Leavit, J.M., Mahoney, A.R., Opie, M., Oshima, T.. and Sanguya, J. (2013). *The meaning of ice. People and Sea Ice in Three Arctic Communities*. International Polar Institute.

Haraway, D. (1988). Situated knowledges: The science question in feminism and the privilege of partial perspective. *Feminist Studies*, 14(3), 575–599.

Harding, S. (2015). *Objectivity and diversity*. University of Chicago Press.

Hauser, D.D., Laidre, K.L., Stafford, K.M., Stern, H.L., Suydam, R.S., and Richard, P.R. (2017). Decadal shifts in autumn migration timing by Pacific Arctic beluga whales are related to delayed annual sea ice formation. *Global Change Biology*, 23 (6), 2206–2217.

ICCI. (2022). State of the Cryosphere Report: Growing losses, growing impacts. International Cryosphere Climate Initiative (ICCI). https://iccinet.org/statecryo22/.

Ingvaldsen, R.B., Assmann, K.M., Primicerio, R., Fossheim, M., Polyakov, I.V., and Dolgov, A.V. (2021). Physical manifestations and ecological implications of Arctic Atlantification. *Nature Reviews Earth & Environment*, 2(12), 874–889.

IPCC. (2021). *Climate Change 2021: The physical science basis. Contribution of Working Group I Contribution to the Sixth Assessment Report of the Intergovernmental Panel on Climate Change*Cambridge University Press.

Ivanova, N., Pedersen, L.T., Tonboe, R.T., Kern, S., Heygster, G., Lavergne, T., Sørensen, A., Saldo, R., Dybkjær, G., Brucker, L., and Shokr, M. (2015). Intercomparison and evaluation of sea ice algorithms: towards further identification of challenges and optimal approach using passive microwave observations, *The Cryosphere*, 9, 1797–1817.

Jasanoff, S. (ed.) (2004a) *States of knowledge: The co-production of science and the social order*. Routledge.

Jasanoff, S. (2004b). Heaven and earth: The politics of environmental images. In S. Jasanoff and M.L. Martello (Eds.) *Earthly politics: Local and global in environmental governance* (pp. 31–52). MIT Press.

Lahsen, M., and Turnhout, E. (2021). How norms, needs, and power in science obstruct transformations towards sustainability. *Environmental Research Letters*, 16 (2), 025008.

Lasserre, F., and Faury, O. (Eds.) (2019). *Arctic shipping: Climate change, commercial traffic and port development*. Routledge.

Latour, B., and Woolgar, S. (1979). *Laboratory life: The construction of scientific facts*, 2nd ed. Princeton University Press.

Maykut, G.A. (1986). The surface heat and mass balance. In *The geophysics of sea ice* (pp. 395–463). Springer US.

Meier, W., and Markus, T. (2015). Remote sensing of sea ice. In M. Tedesco (Ed.) *Remote sensing of the cryosphere* (pp. 248–272). Wiley Blackwell.

Meier, W.N., and Stewart, J.S. (2019). Assessing uncertainties in sea-ice extent climate indicators. *Environmental Research Letters*, 14(3), 035005.

Nagel, T. (1986). *The view from nowhere*. Oxford University Press.

Notz, D., and Stroeve, J. (2018). The trajectory towards a seasonally ice-free Arctic Ocean. *Current Climate Change Reports*, 4(4), 407–416.

Parkinson, C.L., and Kellogg, W.W. (1979). Arctic sea ice decay simulated for a CO_2-induced temperature rise. *Climatic Change*, 2(2), 149–162.

Parks, L. (2005). *Cultures in orbit: Satellites and the televisual*. Duke University Press.

Pedersen, L., Tonboe, R., Kern, S., Lavergne, T., Ivanova, N., and Heygster, G. (2017). Sea ice observations. In T. Carrieres, M. Buehner, J. Lemieux, and L. Toudal Pedersen (Eds.) *Sea ice analysis and forecasting: Towards an increased reliance on automated prediction systems* (pp. 10–50).. Cambridge University Press.

Petrick, S., Riemann-Campe, K., Hoog, S., Growitsch, C., Schwind, H., Gerdes, R., and Rehdanz, K. (2017). Climate change, future Arctic Sea ice, and the competitiveness of European Arctic offshore oil and gas production on world markets. *Ambio*, 46, 410–422.

Polyakov, I.V., Ingvaldsen, R.B., Pnyushkov, A.V., Bhatt, U.S., Francis, J.A., Janout, M., Kwok, R., and Skagseth, Ø. (2023). Fluctuating Atlantic inflows modulate Arctic atlantification. *Science*, 381(6661), 972–979.

Schipper, E.L.F., Dubash, N.K., and Mulugetta, Y. (2021). Climate change research and the search for solutions: rethinking interdisciplinarity. *Climatic Change*, 168(3–4), 18.

Seddon, L.J. (2022). *Measurement, knowledge, and representation: A sociological study of Arctic Sea ice science*. (PhD Thesis). Durham University.

Steinberg, P., and Kristoffersen, B. (2017). 'The ice edge is lost … nature moved it': Mapping ice as state practice in the Canadian and Norwegian North. *Transactions of the Institute of British Geographers*, 42(4), 625–641.

Steinberg, P., Kristoffersen, B., and Shake, K.L. (2020). Edges and flows: Exploring legal materialities and biophysical politics of sea ice. In I. Braverman, and E.R. Johnson (Eds.) *Blue legalities: The life and law of the seas*. Duke University Press.

Tedesco, M. (2015). *Remote sensing of the cryosphere*. Wiley Blackwell.

WMO. (2022). WMO statement on the state of the global climate in 2021. WMO-No. 1290. [Online]. https://library.wmo.int/doc_num.php?explnum_id=11178.

Wormbs, N. (2013). In M. Christensen, A.E. Nilsson, and N. Wormbs (Eds.). *Media and the politics of Arctic climate change: When the ice breaks*. Springer.

9 Marvelous Deposits

The Names and Narratives of Two Hydrothermal Vent Systems

Tirza Meyer and Susanna Lidström

Introduction

In January 2024 the Norwegian parliament voted to allow exploration of the hydrothermal vents located within the country's national waters for possible commercial mining (Meld. St. 25 (2022–2023); Innst. 162 S (2023–2024), 2024; Bryant, 2024). The decision was controversial, as calls for protection of hydrothermal vent systems have increased in recent years and many countries now support a moratorium or pause on deep-seabed mining in international waters so that environmental risks can be further researched and assessed (McVeigh, 2023). To some extent, however, Norway's decision can be considered a win for mining opponents. The vote allowed only exploration, not exploitation, which was an option that was also on the table. In this sense, the decision slows down Norway's progress towards its self-proclaimed aim to become the first nation to extract deep-seabed minerals on a commercial scale (Meld. St. 25 (2022–2023).

Hydrothermal vents are geological features on the deep seafloor that resemble underwater chimneys and vent hot, mineral-rich water from the Earth's crust into the sea. Their discovery along the Galapagos rift in the Pacific Ocean in 1977 was one of the most important oceanographic finds of the twentieth century. It spurred the search for similar sites elsewhere; in subsequent years active and inactive hydrothermal vents have been found all over the world (Ballard, 1977). Around these hydrothermal vents and chimneys ecosystems form that are sustained, not by photosynthesis, but by chemosynthesis, a previously unknown form of life that uses chemicals instead of sunlight to make food. In the aftermath of this discovery, many new scientific theories were developed, such as the idea that vent systems in the deep ocean could potentially provide new clues to the origin of life on Earth (see Martin et al., 2008). Today, scientists continue to discover and identify new species inhabiting vent systems and much is still unknown about the ecosystems and organisms that thrive in these environments (see for example Ramirez-Llodra et al., 2023).[1]

With their fantastic structures and, to us, alien life forms, hydrothermal vents are routinely described as places of marvel and wonder. At the same time, they are also viewed as places brimming with exploitable resources.

DOI: 10.4324/9781003478805-9

Their unique forms of life may hold genetic resources that could be turned into pharmaceuticals, for example, or their vent fluids, which contain minerals that form significant deposits, could be targeted for mining (Koschinsky et al., 2018; McCarthy, 2020).[2] It is possible for the same vent to be viewed as extraordinary and fragile on the one hand and at the same time as a potential site for resource extraction on the other.

Hydrothermal vents are generally named by the scientists who discover them. In this chapter, we consider such naming practices and accompanying descriptions as a way of mediating the ocean and, in particular, its deep-seabed features. Through two case studies we discuss how the naming and characterization of these underwater features shape perceptions of the nature of the vents as well as their relationship to people. We suggest that over time, vent names and descriptions have come to contribute significantly to the incorporation of hydrothermal vents and their ecosystems and minerals into an extractivist narrative and set of related practices that form the predominant way of framing human-ocean relations (Ranganathan, 2019a; see also Lidström, 2023).

We center our analysis on a comparison between the initial descriptions of the hydrothermal vents discovered by the Galapagos expedition in 1977 with the more recent naming and characterization of vent systems discovered on the Arctic Mid-Ocean Ridge, part of the Norwegian Continental Shelf, in the 2000s. The Arctic Mid-Ocean Ridge lies between Greenland, Iceland, and Norway and is, therefore, located in the area which in this book is called the High North Atlantic. Our comparison aims to show how views of hydrothermal vents have changed over time, from an initial focus on their extraordinariness to a perspective which is increasingly resource and exploitation-oriented. We also want to draw attention to differences in the ways in which vent systems are described and regulated, depending on where they are located and under which governance regime they fall. In seabed areas that lay beyond national jurisdiction, resources are governed under the principle of the common heritage of humankind (see Ranganathan, 2019b). In national waters, national rules apply, in our case determined by the Norwegian government.

Through the naming of a hydrothermal vent, a certain type of environment comes into view. A vent's name may inspire subsequent ideas for how to engage with the vent structure, its fauna and minerals. Thus, the naming of vents, we aim to show, is a practice of environing the deep seabed. Historians of technology and the environment Nina Wormbs and Sverker Sörlin have developed the concept of "environing technologies" (Sörlin and Wormbs, 2018) which we use to describe what happens when vent systems receive names and are brought to attention first as places of wonder and later as resources. Furthermore, comparing how vents close to the Ecuadorian Galapagos have been described with those of the vent systems in the High North Atlantic provides a parallel to the chapter in this book by Melody Jue where she compares deep- and cold-water corals with corals in warmer regions.

The Naming of Hydrothermal Vents

Hydrothermal vents occur on the seafloor where tectonic plates move apart, causing seawater to seep down beneath the Earth's crust. As the water heats, it dissolves minerals and causes pressure to build, resulting in mineral-rich water emerging from the seabed in hot springs, or geysers. As the hot water mixes with the cold, the minerals solidify, causing the formation of physical vent structures rising from the seabed. These structures growing out of the Earth's crust are also referred to as white or black "smokers," depending on the color produced by different minerals. Because the vents are the result of ongoing, dynamic movements of the tectonic plates, once they are pushed from the spreading center and the area around them solidifies, they change from active to inactive. The ecosystems of inactive vents usually become extinct, as other vents are typically too far away for species to relocate. This means that many active vent ecosystems are made up of unique, endemic communities (Tunnicliffe, 1992).

Since 1987, some efforts have been made by the International Hydrographic Organization to develop a common standard for the naming of subsea features beyond national jurisdiction (*Standardization of Undersea Feature Names (Guidelines Proposal Form Terminology)*, 2019). However, there are no rules for smaller structures like vent systems or underwater volcanoes. This means that the scientists who discover such features usually decide what to call them on a case-by-case basis. This has led to a variety of naming traditions in national and international waters.

The most reliable source for the names of hydrothermal vent fields is the InterRidge Database, which includes information on a vent's name, its year of discovery, the scientific leader behind its discovery, its geographical coordinates and the name of the vehicle used to identify it. The database does not give a detailed account of how names are chosen but simply provides "the international standard for all known sites of submarine hydrothermal activity which can be updated simply by submitting an electronic message to the InterRidge Office" (Szafranski and Beaulieu, 2020). Stefan Helmreich has commented on the disorderly way of ordering scientific findings within InterRidge, stating that "[w]hat is notable in InterRidge is scientists' move into the realm of do-it-yourself policy, a mode of action seeking to place scientific work in a social context" (Helmreich, 2009, p. 248). In addition to InterRidge, the "U.S: Board on Geographic Names (BGN) Advisory Committee on Undersea Features (ACUF)" (Geographic Names Server), established in 1963, collects both national and international feature place names, including smaller features like vents and reefs, in the so-called *VLIMA Marine Gazetteer* (Claus et al., 2010; Marineregions.org).[3] However, the *VLIMA* list is not exhaustive. Many vent systems that are frequently discussed in scientific publications cannot be found there. The naming of hydrothermal vents, in other words, is semi-regulated and operates in different legal zones and layers of meaning. In general, it appears that the name given to the vent by the scientists who discover it is often the one that sticks, especially in areas beyond national jurisdiction.

Garden of Eden Versus Treasure Chests: Hydrothermal Vents from the Galapagos Rift to the Norwegian Continental Shelf

Hydrothermal vents were first discovered close to the Galapagos Islands in 1977 by an expedition led by Robert Ballard. Ballard reported the "major oceanographic find" later that same year in the journal *Oceanus*, describing the surprising and seemingly otherworldly nature of this previously unknown feature of the seafloor, where life thrived far from sunlight:

> When they reached their target coordinates, Alvin and its three-man crew entered another world. Coming out of small cracks cutting across the lava terrain was warm, shimmering water that quickly turned a cloudy blue as manganese and other chemicals in solution began to precipitate out of the warm water and were deposited on the lava surface, where they formed a brown stain. But even more interesting was the presence of a dense biological community living in and around the active vents. The animals were large, particularly the white clams (up to 30 centimeters or 12 inches). This basis of life was only 50 meters across and totally different from that of the surrounding area. What were the organisms eating? They were living on solid rock in total darkness. (Ballard, 1977, p. 40)

Another team member, geo-archaeologist Tjeerd Hendrik van Andel from Stanford University, described the find in his diary as that of

> a small oasis, perhaps an acre, sharply defined, with coral gardens, pink and gold anemones, white crabs in great variety and profusion, yellow, brown, liver-colored fish, medusoid large clumps of some kind of mussel ten inches long, crevices filled with their huge bleached shells looking from afar like rims of snow in a boulder field. (Oreskes, 2003, p. 722)

He summarized the vent ecosystem as a "little paradise" (Oreskes, 2003, p. 722). The names subsequently given to the vent sites on the Galapagos rift reflect the unexpected abundance of life discovered. The "Garden of Eden," as well as "Rose Garden," "Dandelion patch," "Oyster bed," and "Clambake"[4] highlight the presence and forms of life at each vent, rather than their mineral or geological features. Ballard's report is accompanied by a host of photos of vent fauna taken from the submersible Alvin, including close-ups of crabs, tubeworms, an octopus, fish, as well as unknown species. Ballard emphasizes the vents' uniqueness, noting that "no two were alike" (Ballard, 1977, p. 41).

Scientists had previously posited that hot water emerging from beneath the crust was likely to create mineral deposits. The discovery of such deposits was, therefore, less surprising than the presence of life. While Ballard notes that the

1977 discovery "will likely have a major effect on all disciplines of oceanography" (Ballard, 1977, p. 44), he makes no mention of commercial interests being associated with the vents.

While more vents were discovered across the Earth's oceans in the years following the Galapagos expedition, for a long time it was thought that hydrothermal activity was not to be expected on the "ultra-slow" spreading ridges, like those in the North Atlantic and Bering Sea (see Dick et al., 2003; see also Snow and Edmonds, 2007, p. 96). Attempts to show that this belief was mistaken proved successful when a vent field with active "black smokers" was discovered on the Arctic Mid-Ocean Ridge during a Norwegian-led research cruise in 2008.[5] Geodynamical analysis conducted over a decade had enabled scientists to narrow in on a section where the Mohn's Ridge from the south-west meets the Knipovich ridge

coming from the north. One day at sea in July 2008, the researchers deployed the Remotely Operated Vehicle (ROV) *Argus Bathysaurus* from their vessel *G.O. Sars*, which was positioned over the ridge section in question. At just over 2000 meters depth, the ROV encountered towering chimneys emitting what looked like black smoke. Crew member Marvin Lilley later described the chimneys as "mostly black and covered with white mats of bacteria feasting on the minerals emitted by the vents" (ScienceDaily, 2008). The scientists also found less active chimneys "mottled red as a result of iron oxidization" (ScienceDaily, 2008). They noted the "huge mound of sulfide minerals" beneath one of the vents and described the deposit as "about 825 feet in diameter at its base and about 300 feet across on the top" and stated that it "might turn out to be the largest such deposit seen on the seafloor" (ScienceDaily, 2008). As on the Galapagos cruise, during their dive the scientists encountered what they saw as a unique habitat. They found out later that many of the species living there were endemic to this specific vent system (Pedersen et al., 2010).[6]

The first official report of vent discovered on the Arctic Mid-Ocean Ridge was published in 2008 in the international news outlet of the InterRidge Database (Pedersen, 2008). Expedition leader and marine geologist Rolf Birger Pedersen described the new find in the short news section: "The first leg of this cruise succeeded in locating a new black smoker vent field at ~2400m water depth" (Pedersen, 2008). The news segment included a photo of a "chimney" with two branches. Also in the photo is the Bathyrasaurus' hydraulic arm in the process of sampling fluids. Pedersen's short note also revealed that the discovered vent had been given a name: Loki's Castle, or *Lokeslottet* in Norwegian. Loki, or Loke, is a character from Norse mythology who is often described as an ambivalent trickster who cannot be trusted. The online news site *Science Daily* stated that "The field has been named Loki's Castle partly because the small chimneys at the site looked like a fantasy castle to the scientists" (ScienceDaily, 2008), referring to a press release from the University of Bergen about the discovery that also said that Loki "was an appropriate name for a field that was so difficult to locate" (ScienceDaily, 2008).

After the initial discovery, the Norwegian scientists returned to the site in 2009 and again in 2010 for further research. This work culminated in the first associated scientific publication titled "Discovery of a black smoker vent field and vent fauna at the Arctic Mid-Ocean Ridge," in which Pedersen described the sheer size of the "deposit" in more detail (Pedersen et al., 2010):

> The field is composed of four active black smoker chimneys, up to 13 m tall, at the top of a mound of hydrothermal sulphide deposits. Venting of 310–320°C black smoker fluids occurs at two sites that are around 150 m apart. (Pedersen et al., 2010, p. 2)

The article also includes photos of a selection of "characteristic invertebrates" (Pedersen et al., 2010, p. 4), including tubeworms waving in deep water in front of what looks like white foam, but is in reality the microbial mats that the scientist encountered when they first discovered the vent system. Pedersen's article closes with the stated hope that the discovery of this vent could lead to "a new opportunity to advance our understanding of the migration of vent fauna and interactions between different chemosynthetic deep-sea environments" (Pedersen et al., 2010, p. 4).

Loki's Castle is not the only vent system that has been discovered along the Arctic Mid-Ocean Ridge, although it was the first discovery of an active vent field. Relatively close by, on Mohn's Ridge, an inactive vent system had been discovered in 2002 and named Mohn's Treasure.[7] Other vent systems located on the Norwegian Continental Shelf have been named following a similar logic. Gnitahei and Fåvne are named after a dragon who hovers over and protects a golden treasure. Soria Moria was named after a castle in a Norwegian fairy tale (Peterson, 2013)[8] where a poor young man goes on a transformative journey that lifts him out of poverty. The young man embarks on a sea voyage, finds land and discovers a castle: "Far, far away something that looked like a little star was shining. He got closer and closer, and then further away, east of the sun and west of the moon, Soria Moria Palace shimmered like gold." Many of the other mythological names given to vent systems likewise indicate some sort of treasure or place of high value. While Norse mythology is much older than the Norwegian nation, and is connected to a larger cultural area, these naming practices can be seen as signaling that these discoveries (and resources) naturally or historically belong to the Norwegian people who lived here long before the drawing of any national borders.

Norway has a similar tradition of drawing on mythology in its naming of oil fields. In the 1980s, the Norwegian government, along with the National Council for Language (Språkrådet) and the Norwegian Petroleum Directorate (after January 1, 2024 the Norwegian Offshore Directorate), compiled a list of suitable names to use for newly-discovered oil fields. Most of the names on this list were taken from Norse mythology. In 2011, this naming policy was changed and a larger variety of names were added to the list. The final naming decision is still taken by the Directorate.[9] There are now names in use that are inspired by seabirds and coastal place names. However, to date, names from Norse mythology continue to dominate (see Hovednak, 2018; see also Helgesen, 2011). Several of the names given to the hydrothermal vents discovered by Norwegian scientists seem to reflect this tradition.

A few years after the discovery of Loki's Castle, Pedersen suggested that the area around Loki's Castle be protected and turned into a marine park. The proposal was published as a news item by the University of Bergen in August 2013 with a quote from Pederson stating that "It is our opinion that this area is so unique that it should be preserved." Pederson also pointed out that "[w]e are talking about very vulnerable environments" and that "research also needs to create more knowledge about the wildlife in the area" (Wehus and Drøne, 2013). To date, the Norwegian government has not established a conservation area around this vent system, although the scientific call for protection persists (Eilertsen et al., 2024).

A 2016–2017 report clarifies the Norwegian government's reasoning around the mineral deposits at Loki's Castle. The report recognizes the uniqueness of the fauna, noting that "Loki's Castle, which was discovered in 2008, and which contains the largest mineral deposits that have so far been detected in the Norwegian Sea, was the first discovery of species that are particularly adapted to hot springs in our deep-sea areas, species that are also endemic to the area."[10] However, it then proceeds to state that "[i]t is important to examine the areas both before, during and after any activity. Preliminary investigations are also needed in surrounding areas to map the possibility of recolonization of the fauna."[11] Instead of responding to suggestions for protection, the government states the need for plans of recolonization after any disturbance associated with mineral removal. In other words, the report seems to assume that minerals should and will be mined, as opposed to being protected in their current state.

To date, the protection status of Loki's Castle remains unclear. The *Marine Gazetteer* has added Loki's Castle as a marine protected area (MPA) but cites no reliable source (Marine Regions Loki's Castle (Marine Protected Area (MPA)). In 2021, the Norwegian government published a report where the unique environment around the vent system was stressed once again (Meld. St. 29 (2020–2021), 2021, pp. 13–15). At the same time, one of the report's sections ends with the observation that "[i]mportant mineral deposits have been found along the Mid-Atlantic Ridge, and parts of it may therefore be of interest for mineral extraction from the seabed" (Meld. St. 29 (2020–2021), 2021, p. 14; see also Meld. St. 25 (2022–2023)).

Although there are currently no official plans to mine Loki's Castle, there is a mining company claiming its name. In 2019, the Norwegian metals company "Loke minerals" was founded, a direct reference to the Loki's Castle vent field (Lokemm.com). When asked in an interview whether or not the company was planning to mine sulphide crusts on hydrothermal vents, Wenche R. Helleland, the head of environment and sustainability, said that "So far, we are not working with technology that targets this. On the

Norwegian Continental Shelf, there is talk of extracting manganese crusts" (Svendsen, 2023). The company is awaiting approval of deep-seabed mining on the Norwegian Continental Shelf (Lokemm.com).

When scientists describe and name hydrothermal vents, their actions and decisions have implications for how those environments are perceived and understood outside the scientific realm. Names and their accompanying descriptions function as "environing technologies" that help transform the deep seafloor in general, and hydrothermal vents in particular, from wilderness areas beyond social realms and public interest into "environments" that are part of human imaginations and spheres of influence (Sörlin and Wormbs, 2018). Together with the physical technologies necessary to explore the deep sea and the scientific theories that describe them, names and characterizations of hydrothermal vents, their ecosystems and minerals turn these deep-sea features not only into known natures, but also equip them with cultural, political, economic and commercial valuations and relevance.

Viewed as an environing technology, scientific naming and characterizations emerge as a central step in the environing of deep-sea hydrothermal vents. The words used to describe newly-discovered vents inevitably carry connotations, relating them to human histories, experiences and viewpoints. When the scientists on board the submersible Alvin watched as a vent community came into view for the first time in recorded human history, when describing what they saw they used words like "oasis" and "coral gardens," and named the vent systems after the organism they found there. The name "Oyster Patch" was the result of a misidentification, indicative of the fact that there was not "a single biologist on the entire cruise—only geologists, geophysicist, chemists, geochemists, physicists, and a science writer" (Ballard, 1977, p. 41). Such an oversight speaks to the fact that, while the existence of hot springs had been predicted by geologists, the presence of life around them came as a complete surprise. Considering the composition of expertise among the crew, it is perhaps surprising that in their report, the minerals venting from the structures, described simply as "manganese and other chemicals," were mentioned only in passing (Ballard, 1977, p. 40). This neglect indicates that right from the start the biological discovery took center stage, even among the geophysicists.

The Norwegian-led crew that went out to look for vent activity on the Arctic Mid-Ocean Ridge in 2008 was already familiar with the biodiversity of vent systems on fast spreading subduction zones. Unlike the Galapagos crew of 30 years earlier, the scientists on the Sars focused much more on the "deposit" aspect of their find. For example, crew member Marvin Lilley talked about the chimneys "sitting on the huge deposit" (ScienceDaily, 2008). Calling the site a "castle," with its connections to rocks and mountains, highlights the fact that the vent is teeming with minerals rather than previously undiscovered biological life (although the presence of biodiversity is mentioned as well).

Words like “chimney” and “smoker”—frequently used to describe hydrothermal vents—associate these deep-sea areas with urban manufacturing landscapes. “Huge chimneys,” “like factories,” the “smokers,” and even the color black, the heat and the frequent references to the sheer size of the vents, all reference the image of a typical industrial landscape and can be read as suggesting that hydrothermal vents are environments of industrial activity put in place to work for us. The fact that there is now a “castle” that “sits on the deposits” suggests that there is a treasure waiting to be recovered.

Marvelous Deposits

We have seen how two different ways of describing and naming vent systems help to transform hydrothermal vents from places of unknown wilderness to familiar features in contemporary environmental imaginations. What kind of environments emerge from the different ways of describing and naming vents? The rich descriptions of life around the vents in the Galapagos case and the names taken directly from either terrestrial “nature,” “gardens,” or the organisms that were observed to be living there, evoke an environment of life and richness that is unique. It also gives the organisms living around the vents an agency or a position of ownership to the places they inhabit. The vents are their Garden of Eden. The names of the Norwegian features, on the other hand, emphasize different characteristics, portraying the vent sites either as inanimate objects, like castles sitting on treasures, or mythical creatures. The names evoke classic fairytale-inspired narratives of finding the treasure, fighting the dragon, and claiming the gold.

We suggest that the naming practices and rhetoric surrounding the exploration of vents in Norwegian waters demonstrates how hydrothermal vents have been gradually incorporated into an extractivist narrative. This framing is not unique to hydrothermal vents; it is a fundamental narrative describing human-Earth relationships, not least when it comes to contemporary views of the deep ocean (see Ranganathan, 2019a; see also Lidström, 2023). Legal scholar Surabhi Ranganathan has shown how international ocean law is shaped by an extractivist imaginary spurred by the presence of potentially exploitable minerals on the deep seabed, so that the United Nations Law of the Sea (UNCLOS), negotiated for decades before being adopted in 1982, “makes the extraction of resources a normative obligation: not just something permitted, or to be regulated, but also something to be promoted” (Ranganathan, 2022, p. 32).

In other words, when the first hydrothermal vents were discovered there was already a pre-existing narrative characterizing deep-sea minerals as an economic resource. Unsurprisingly, vent minerals quickly became part of this narrative. In the Norwegian context this process was especially rapid. For example, in the same year that Loki’s Castle was discovered, the now defunct mining firm Nautilus Minerals identified several deposits within the jurisdiction of Papua New Guinea with the intention of mining them (see Massoth et al., 2008).[12] The extractivist narrative underlying the discovery of the vent systems on the Norwegian Continental Shelf was not only apparent in government reports, which explicitly referred to commercial exploitation of the vents, but was also, in more subtle ways, in evidence in the scientific reports compiled at the time, which frequently used words like "deposit" and referred to vents as "treasures."

In addition to the extractive interests in vent minerals, there is also a potentially competing commercial interest in the biological communities found around the vents. Due to their unique life forms, vent organisms are believed to have high potential value as so-called marine genetic resources, or MGRs, of interest in the development of pharmaceuticals and other biotechnological products (see for example Broggiato, 2013). While viewing vents as possible sources for commercial MGRs also turns them into a potential resource, the practices involved in extraction and exploitation are significantly different and likely to be less destructive than commercial mining. Rather than a sweeping removal of entire vent structures, biological extraction would more likely consist of small-scale sampling. Moreover, the removal of minerals and associated destruction of vent habitats would likely remove the possibility of identifying medically or otherwise commercially interesting MGRs, resulting in two competing extractive interests surrounding the vents.

From a legal perspective, MGRs are fundamentally different from minerals. According to UNCLOS, in waters beyond national jurisdiction the distinction between geology and biology is "sharply binary: marine life is placed in the high seas, and minerals under the seabed regime." The Norwegian government seems to favor the more destructive prospect of mineral extraction and largely ignores the idea to protect the area around Loki's Castle. Instead of seeking to preserve the unique, endemic species living in and around the vent site it instead maintains a consistent focus on the inorganic resources located underneath the vent ecosystems. A recent scientific report describing for the first time the fauna of the Norwegian vents provides a counter narrative highlighting the uniqueness and diversity of the species living at Loki's Castle (Eilertsen et al., 2024).

Through analysis of naming practices and the narratives surrounding them, this chapter has aimed to show the tensions between contrasting and conflicting perspectives evident in the discoveries of deep-sea hydrothermal vents. Conservationist and commercial interests clearly compete against and interact with each other, as in the claims that deep-sea minerals are essential sources of materials for electrification. Vent system naming, we argue, is a way of mediating specific sites in the ocean. This form of mediation has consequences for how deep-sea features are viewed now and may be treated in the future. Comparing the initial discovery of hydrothermal vent ecosystems in the 1970s to recent exploration of vents in Norwegian waters indicates a movement towards an increasing "resourcification" of the vents. This is not surprising. Instead, it shows how the vents and their associated ecosystems have quickly and seamlessly become part of an overarching, and legally underpinned, extractivist imaginary of human-ocean relations. The part of the

High North Atlantic between Greenland, Iceland and Norway has a long history of industrial offshore activity for oil extraction. In contrast, the site of the first vent discovery, the Galapagos, has a long history of recognized biodiversity and scientific values, with conservation efforts stretching back to Darwin's *On the Origin of Species* (1859). The location of the vents we compare is this study follow a similar interpretative trajectory which may have influenced the choice of names as well as ideas of how to treat and manage these discoveries.

Scientists (and others) still consider hydrothermal vents and their inhabitants as marvels, just as the original Galapagos crew did. As the gaze of governments and industry focus increasingly on the "deposits" that sit underneath these features, scientific, legal, commercial and conservationist narratives have come to compete with and seep into each other, making the vents complex and contested sites in marine environmental imaginations. Scientific reports, perhaps unintentionally, often perpetuate extractivist views by adopting a rhetoric centered on the mineral "deposits" as integrated parts of entire vent structures. As no mining has yet commenced or been approved, vents are currently in an in-between state: they are places of both marvel and extractive potential. As long as there is no mining activity at vent sites, it is likely that these distinctive micro-environments will retain their status as "marvelous deposits."

Notes

1 The communities inhabiting vent systems are called extremophiles because, from an anthropogenic viewpoint, they can thrive in extreme conditions. See for example Marrs and Barry (1997).
2 For the recent status of bioprospecting, see Stokke et al. (2020, p. 249).
3 This database is different from the Gazetteer of Undersea Feature Names hosted by the British Oceanographic Data Centre (BODC) (GEBCO Undersea Feature Names, n.d.).
4 The full list of names for the vent fields in Rose Garden can be found at https://vents-data.interridge.org/ventfield/rose-garden
5 For the first published account of the discovery see Rolf Pederson, "Norway," in InterRidge News, Volume 17, 2008, p. 37. Available at https://www.interridge.org/wp-content/uploads/2022/01/IRNewsVol17-1_2008.pdf. Since that time many more vent fields have been discovered in this area.
6 Pedersen et al. was the first publication to describe the discovery of these new species. Years later, new species continue to be found at the vent site (see, for example, Hamstad, 2021).
7 This map shows the locations of the known hydrothermal sulphide deposits identified along the Mohn's ridge system. It also shows the Loki's Castle deposit just to the north-east (Brekke et al., 2021, p. 17) (see also Ramirez-Llodra et al., 2020).
8 Other vent systems have been given mythological names like Lokeslottet, Fåvne, Ægirs kilde, Soria Moria, Trollvegge, Perle and Bruse and Syv Søstre. For an overview of all of the hydrothermal fields on the Norwegian shelf and their geological and biological compositions see Pedersen et al. (n.d., pp. 37–41).
9 For the rules concerning the naming of oil fields, see Forskrift Til Lov Om Petroleumsvirksomhet—Kapittel 10. Alminnelige Bestemmelser—Lovdata (n.d., see also Hovland, 2011).

10 Meld. St. 35 (2016–2017). Oppdatering av forvaltningsplanen for Norskehavet. p. 24. Available at: https://www.regjeringen.no/no/dokumenter/meld.-st.-35-20162017/id2547988/?ch=3. English translation. Original Norwegian text: "*Lokeslottet,* som ble oppdaget i 2008, og som innholder de største mineralavsetningene som til nå er påvist i Norskehavet, var det første funnet av arter som er spesielt tilpasset varme gassoppkommer i våre dyphavsområder, arter som også er endemiske for området (arter som bare finnes her)."

11 Meld. St. 35 (2016–2017). Oppdatering av forvaltningsplanen for Norskehavet. p. 24. Available at: https://www.regjeringen.no/no/dokumenter/meld.-st.-35-20162017/id2547988/?ch=3. Original Norwegian text: "Områder rundt geotermiske oppkommer har sjeldne naturtyper og helt spesielle organismer og arter. Det er viktig å undersøke områdene både før, under og etter eventuell aktivitet. Også i omkringliggende områder er det nødvendig med forundersøkelser for å kartlegge muligheten for rekolonisering av faunaen."

12 More research needs to be done on the difference between naming practices in international waters, those in waters of a foreign state and national waters. Is there a possible inclination to give names that signal ownership only when the vents are in the national jurisdiction of a country that is also home to some of the crew on the research vessels? A good place to start looking for an answer would be the InterRidge Database where all names are listed. On Interridge see https://vents-data.interridge.org/ventfield/solwara-17.

References

Ballard, R.D. (1977). Notes on a major oceanographic find. *Oceanus*, 20(3), 35–44.

Brekke, H., Stenløkk, J., Eriksen, S.H., Bjørnstad, A., Sandstå, N.R., Solvi, K., and Schiager, P. (2021, August 1). Deep sea minerals on the Norwegian Continental Shelf: Developments in exploration [Norwegian Petroleum Directorate]. Geological Society of Norway's winter meeting in January 2021.

Broggiato, A. (2013). Exploration and exploitation of marine genetic resources in areas beyond national Jurisdiction and Environmental Impact Assessment. *European Journal of Risk Regulation*, 4(2), 247–251.

Bryant, M. (2024, January 9). Norway votes for deep-sea mining despite environmental concerns. *The Guardian*. https://www.theguardian.com/environment/2024/jan/09/norway-set-to-approve-deep-sea-mining-despite-environmental-concerns.

Claus, S., De Hauwere, N., Vanhoorne, B., and Hernandez, F. (2010). The VLIMAR Gazetteer: A standard for marine geographic features. In M. Fichaut*et al.* (Eds.), *International Marine Data and Information Systems Conference (IMDIS 2010) Book of Abstracts* (p. 185).

Dick, H.J.B., Lin, J., and Schouten, H. (2003). An ultraslow-spreading class of ocean ridge. *Nature*, 426(6965), 405–412. doi:10.1038/nature02128.

Eilertsen, M.H., Kongsrud, J.A., Tandberg, A.H.S., Alvestad, T., Budaeva, N., Martell, L.*et al.* (2024). Diversity, habitat endemicity and trophic ecology of the fauna of Loki's Castle vent field on the Arctic Mid-Ocean Ridge. *Scientific Reports*, 14(1), 103. doi:10.1038/s41598-023-46434-z.

Forskrift Til Lov Om Petroleumsvirksomhet Kapittel 10. Alminnelige Bestemmelser—Lovdata, *Pub. L. No. § 79. Navnsetting mv. av petroleumsforekomster.* Retrieved March 5, 2024, from https://lovdata.no/dokument/SF/forskrift/1997-06-27-653/KAPITTEL_12#%C2%A779.

GEBCO Undersea Feature Names. (n.d.). *A digital gazetteer of the names of features on the seafloor* [dataset]. Retrieved March 15, 2024, from https://www.gebco.net/data_and_products/undersea_feature_names/.

Geographic Names Server. (n.d.). *U.S. Board on Geographic Names (BGN) Advisory Committee on Undersea Features (ACUF)* [dataset]. Retrieved March 5, 2024, from https://geonames.nga.mil/geonames/GNSHome/reference.html.

Hamstad, I.M. (2021). *Characterization of benthic communities at Loki's Castle vent field using a photomosaic* [Master thesis, The University of Bergen]. https://bora.uib.no/bora-xmlui/handle/11250/2779706.

Helgesen, O.K. (2011, December 6). Ny navneliste til ingen nytte. *Teknisk Ukeblad.* https://www.tu.no/artikler/ny-navneliste-til-ingen-nytte/247300.

Helmreich, S. (2009). *Alien Ocean: Anthropological Voyages in Microbial Seas.* University of California Press.

Hovednak, M. (2018, December 1). Femit år med norske olje- og gassfelt. *Romsdals Budstikke.*

Hovland, K.M. (2011, April 5). Oljefelt heter ikke Harry Potter. *Teknisk Ukeblad.* https://www.tu.no/artikler/oljefelt-heter-ikke-harry-potter/247667.

Koschinsky, A., Heinrich, L., Boehnke, K., Cohrs, J.C., Markus, T., Shani, M., et al. (2018). Deep-sea mining: Interdisciplinary research on potential environmental, legal, economic, and societal implications. *Integrated Environmental Assessment and Management*, 14(6), 672–691. doi:10.1002/ieam.4071.

Lidström, S. (2023). Ocean literacy and marine environmental narratives: Interpretations of sustainability in the deep sea. *Resilience: A Journal of the Environmental Humanities*, 10(3), 43–68.

Lokemm.com. (n.d.). Deep sea minerals for the green transition. Loke Marine Minerals. Retrieved March 5, 2024, from https://lokemm.com/.

Marine Regions Loki's Castle (Marine Protected Area (MPA)). (n.d.). Retrieved March 5, 2024, from https://marineregions.org/gazetteer.php?p=details&id=32602.

Marineregions.org. (n.d.). *Marine Gazetteer geographic name search* [dataset]. Retrieved March 5, 2024, from https://www.marineregions.org/gazetteer.php?p=search.

Marrs, M.T.M., and Barry L. (1997, April 1). Extremophiles. *Scientific American.* Retrieved March 5, 2024, from https://www.scientificamerican.com/article/extremophiles/.

Martin, W., Baross, J., Kelley, D., and Russell, Mi. J. (2008). Hydrothermal vents and the origin of life. *Nature Reviews Microbiology*, 6, 805–814.

Massoth, G., Puzic, J., Crowhurst, P., White, M., Nakamura, K., Walker, S., and Baker, E. (2008). Regional Venting in the Manus Basin, New Britain Back Arc. *AGU Fall Meeting Abstracts.*

McCarthy, R. (2020). Deep sea rush: With valuable metals on the ocean floor, speculators are circling. *The Baffler*, 54, 114–124.

McVeigh, K. (2023, October 30). UK backs suspension of deep-sea mining in environmental U-turn. *The Guardian.* Retrieved March 5, 2024, from https://www.theguardian.com/environment/2023/oct/30/uk-backs-suspension-of-deep-sea-mining-in-environmental-u-turn.

Meld. St. 25 (2022–2023). (n.d.). Mineralverksemd på norsk kontinentalsokkel – opning av areal og strategi for forvaltning av ressursane [Stortingsmelding]. Olje- og energidepartementet. Retrieved March 5, 2024, from https://www.regjeringen.no/nn/dokumenter/meld.-st.-25-20222023/id2985856/.

Meld. St. 25 (2022–2023), Innst. 162 S (2023–2024). (2024). *Innstilling fra energi- og miljøkomiteen om Mineralverksemd på norsk kontinentalsokkel—opning av areal og strategi for forvaltning av ressursane*. Retrieved March 5, 2024, from https://www.stortinget.no/no/Saker-og-publikasjoner/Publikasjoner/Innstillinger/Stortinget/2023-2024/inns-202324-162s/.

Meld. St. 29 (2020–2021). (2021). Norway's integrated plan for the conservation of areas of special importance for marine biodiversity [Meld. St. 29 (2020–2021)]. *Norwegian Ministry of Climate and Environment*. Retrieved March 5, 2024, from https://www.regjeringen.no/contentassets/cb23f1dc245f43df821fe96a53d66de6/en-gb/pdfs/stm202020210029000engpdfs.pdf.

Oreskes, N. (2003). A context of motivation: US Navy oceanographic research and the discovery of sea-floor hydrothermal vents. *Social Studies of Science*, 33(5), 697–742. doi:10.1177/0306312703335004.

Pedersen, R.B. (2008). Norway. *InterRidge News*, 17, 39.

Pedersen, R.B., Rapp, H.T., Thorseth, I.H., Lilley, M.D., Barriga, F.J.A.S., Baumberger, T. et al. (2010). Discovery of a black smoker vent field and vent fauna at the Arctic Mid-Ocean Ridge. *Nature Communications*, 1, 126. Retrieved March 5, 2024, from doi:10.1038/ncomms1124.

Pedersen, R.B., Rydland Olsen, B., Barreyre, T., Bjerga, A., Denny, A., Heggernes Eilertsen, M. et al. (n.d.). Fagutredning Mineralresursser i Norskehavet Landskapstrekk, Naturtyper og Bentiske Økosystemer. SENTER FOR DYPHAVSFORSKNING UNIVERSITETET I BERGEN. Retrieved March 5, 2024, from https://www.regjeringen.no/contentassets/dbf5144d0fbc42b5a4db5fc7eb4fa312/vedlegg-3.-senter-for-dyphavsforskning-uib.-landskapstrekk-naturtyper-og-bentiske-okosystemer.pdf.

Peterson, I. (2013, October 22). Deepwater Mining in Norway. *Science & Tech*. Retrieved March 5, 2024, from https://www.oceannews.com/news/science-technology/deepwater-mining-in-norway.

Ramirez-Llodra, E., Argentino, C., Baker, M., Boetius, A., Costa, C., Dahle, H. et al. (2023). Hot vents beneath an icy ocean: The Aurora Vent Field, Gakkel Ridge, Revealed. *Oceanography*, 36(1), 6–17.

Ramirez-Llodra, E., Hilario, A., Paulsen, E., Costa, C.V., Bakken, T., Johnsen, G., and Rapp, H.T. (2020). Benthic communities on the Mohn's treasure mound: Implications for management of seabed mining in the Arctic mid-ocean ridge. *Frontiers in Marine Science*, 7. Retrieved March 5, 2024, from https://www.frontiersin.org/articles/10.3389/fmars.2020.00490.

Ranganathan, S. (2019a). Ocean floor grab: International law and the making of an extractive imaginary. *European Journal of International Law*, 30(2), 573–600. Retrieved March 5, 2024, from doi:10.1093/ejil/chz027.

Ranganathan, S. (2019b). The common heritage of mankind: Annotations on a battle. In J. von Bernstorff and P. Dann (Eds.), *The Battle for International Law: South-North Perspectives on the Decolonization Era*. Oxford University Press. Retrieved March 5, 2024, from doi:10.1093/oso/9780198849636.003.0002.

Ranganathan, S. (2022). The vexed liminality of hydrothermal vents: An opportunity to unmake the law of the sea. In I. Braverman (Ed.), *Laws of the Sea: Interdisciplinary Currents*. Routledge.

Ruud, L. C. (2019). Oil as heritage: Toponymies and temporalities on the Norwegian Continental Shelf. *Ethnologia Scandinavica*. Retrieved March 5, 2024, from https://www.academia.edu/40439957/Oil_as_Heritage_Toponymies_and_Temporalities_on_the_Norwegian_Continental_Shelf.

Ruud, L.C. (2022). Oil and Vikings: Temporal alignments within Norwegian petroleum fields. In A. Ekström and S. Bergwik (Eds.), *Times of History, Times of Nature: Temporalization and the Limits of Modern Knowledge*. Berghahn.

ScienceDaily. (2008, July 24). Scientists break record by finding northernmost hydrothermal vent field. ScienceDaily. Retrieved March 5, 2024, from https://www.sciencedaily.com/releases/2008/07/080724153941.htm.

Snow, J. and Edmonds, H. (2007). Ultraslow-spreading ridges: Rapid paradigm changes. *Oceanography*, 20, 90–101. Retrieved March 5, 2024, from doi:10.5670/oceanog.2007.83.

Sörlin, S. and Wormbs, N. (2018). Environing technologies: A theory of making environment. *History and Technology*, 34(2), 101–125. Retrieved March 5, 2024, from doi:10.1080/07341512.2018.1548066.

StandardizationofUnderseaFeature Names(GuidelinesProposalFormTerminology) (B-6; 4.2.0). (2019). Retrieved March 5, 2024, from https://iho.int/en/bathymetric-publications.

Stokke, R., Reeves, E.P., Dahle, H., Fedøy, A.-E., Viflot, T., Lie Onstad, S. et al. (2020). Tailoring hydrothermal vent biodiversity toward improved biodiscovery using a novel in situ enrichment strategy. *Frontiers in Microbiology*, 11. Retrieved March 5, 2024, from https://www.frontiersin.org/journals/microbiology/articles/10.3389/fmicb.2020.00249.

Svendsen, K. (2023, March 21). Gründerne i Loke Marine Minerals har kjøpt letelisenser og sikter mot mineralutvinning på havbunnen. Kystens Næringsliv | Vi Skaper Kysten Sammen! Retrieved March 5, 2024, from https://www.kystens.no/energi/gr-nderne-i-loke-marine-minerals-har-kjopt-letelisenser-og-sikter-mot-mineralutvinning-pa-havbunnen/2-1-1422539.

Szafranski, K., and Beaulieu, S.E. (2020). InterRidge global database of active submarine hydrothermal vent fields. *Version 3.4*. Retrieved March 5, 2024, from https://vents-data.interridge.org/.

Tunnicliffe, V. (1992). Hydrothermal-vent communities of the deep sea. *American Scientist*, 80(4), 336–349.

Wehus, W., and Drøne, S.O. (2013, February 8). A waterworld of volcanos. *UiB News*. Retrieved March 5, 2024, from https://www.uib.no/en/news/36493/waterworld-volcanoes.

10 The Battle of the Seabed

Mining, Mediating and Territorializing the Deep Sea

Jørgen Alnæs

Sounds From the Deep

On February 10, 2023, a line was drawn between the Norwegian capital and the deepest and darkest part of the seabed between Norway and Greenland, on the extreme periphery of Norwegian territory.

The University of Oslo's ceremony hall is one of the most esteemed assembly halls located in the middle of the city, between the royal castle and the parliament. The distinguished neo-classical banquet hall is decorated with monumental wall paintings by Edvard Munch. Above the stage, the famous painting "The Sun" represents the university's fundamental task—enlightenment—expressed as a majestic sunrise. For years, this was the location for the Nobel Peace Prize award ceremony.

On this winter evening, people were gathered in the assembly hall to listen to a premiere performance of the composition "Loki's Castle" (In Norwegian: "Lokeslottet"). Loki is a central god in Norse mythology, but the title of the actual composition referred to a place named after the Norse god's habitation: *Loki's Castle* is one of a few recently discovered underwater hydrothermal chimneys or vents, approximately midway between Norway, Greenland, Iceland, and Spitsbergen. The vents are located at a depth of 2,350 meters on the Mid-Atlantic Ridge and are part of the extended Norwegian continental shelf.

This music sounds "pretty wild," said the chairman of the choir, Knut Olav Åmås, in his opening speech. As a director of the Norwegian Foundation of the Free Speech, an author and former editor, Åmås is a well-known figure in Norway. By Åmås' short talk, the premiere performance of Loki's Castle was framed as highbrow and important. During her own introduction, the composer Ruth Wilhelmine Meyer explained that the music aimed to illustrate the shared connections of humans and creatures found at an altitude of 2,350 meters below the surface of the sea (Aarnes, 2023). She emphasized in the printed program, "We need emotional identification with the ocean to act" (The Norwegian Soloist's Choir, 2023, p. 2).

Both the composition and the concert must be seen in the context of an ongoing public debate on deep-sea mining in Norway. During the beginning

DOI: 10.4324/9781003478805-10

of the 2020's, new ocean industries have entered the Norwegian public debate, and also, as a consequence, new aspects of the ocean and the Norwegian waters. In the same manner as with offshore wind and sea-wave power, the idea of deep-sea mining has resulted in discussions of conservation versus industry. Half a year earlier, at the UN Ocean Conference in Lisbon, the French president Emanuel Macron argued for a legal framework against deep-sea mining. France will never support such industries, he stated (Valo, 2022). Macron's speech was referred to in several Norwegian newspapers and contrasted with the Norwegian government which had initiated a discussion on the topic (Andersson, 2022; Fransson, 2022). Against this background, the concert in Oslo appears to be a subtle argument in favor of conservation.

Still, the concert has to be seen in wider context: both the premiere performance of the composition Loki's Castle, as well as the deep-sea mining debate in Norway, are the result of a huge expansion of the Norwegian continental shelf. Twenty years earlier, neither Loki's Castle nor some of the other underwater hydrothermal vents were part of Norwegian territory. They were undepicted and more or less unnamed, and no states claimed this part of the seabed. The area was a tabula rasa and a part of the Global Commons deep-sea floor known as The Area (In Norwegian: "Området"). Nevertheless, in March 2009, the UN Commission on the Limits of the Continental Shelf accepted a proposition to extend the Norwegian continental shelf, and this part of the seabed was included in Norway's territory (Meld. St. 20 (2019–2020)).

This chapter examines how this previously unidentified no man's land or Global Commons was included—and performed—as a part of Norway. Through two case studies of journalistic texts, the chapter will discuss how the seabed is performed through the texts, how it acquires meaning and significance, but also how it acts as a mirror to confirm or show what is part of and what is not part of the Norwegian discourse and the Norwegian self-image.

Fluid Territoriality

From a terrestrial point of view, borders and sovereignty in the ocean are complicated. While the border between two states' land-based territories usually can be marked with one single line, the state's sovereignty in the sea is divided into zones with different degrees of jurisdiction (see Figure 10.1). Along the coast, in *the internal waters*, the coastal state has the same sovereignty as on land. In *the territorial sea, the contiguous zone* and *the economic zone*, other nations have different rights, but the coastal state still controls the marine resources from the surface of the sea to the seabed, from fish and mammals to seaweed, petroleum, and minerals.

While the exclusive economic zone stretches 200 nautical miles from the coast, the continental shelf can, after negotiations, be expanded beyond this 200-nautical-mile zone (United Nations 1982). In such cases, the coastal state can control extraction from the seabed, but not from the waters above it. This is the case in the Norwegian Sea, which is the marginal sea between Norway

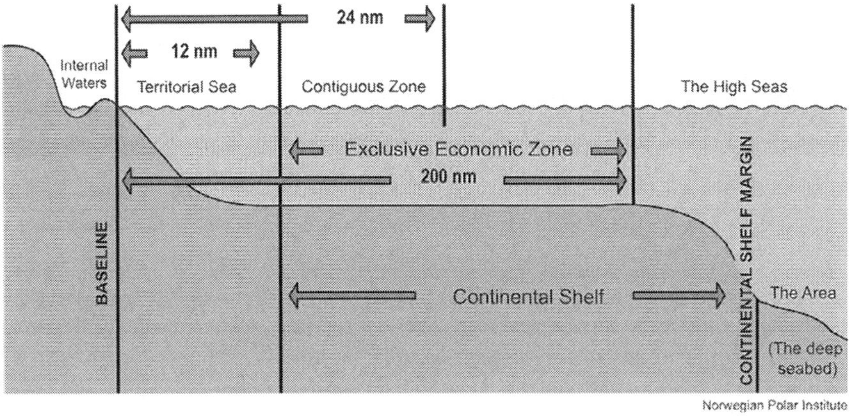

Figure 10.1 Maritime zones defined by United Nations Convention on the Law of the Sea (UNCLOS)

Source: Illustration by Gunnar Sander and Audun Igesund / Norsk Polarisntitutt (Norwegian Polar Institute).

and Greenland. Northeast of Iceland, almost in the middle of the Norwegian Sea, there is an area that is outside any state's 200-mile zone. This is *open sea*, which means that none of the three countries have jurisdiction there. Because of the area's shape on a map, the area has been given the name The Banana Hole.[1] The Banana Hole has been characterized by various disputes concerning fishing. Until 2009, *the seabed* beneath The Banana Hole was also outside the jurisdiction of any single state.

The idea of the ocean depths being a common property or Global Commons is often traced back to the Maltese Ambassador to the UN Arvid Pardo's speech at the UN in 1967 (Pardo, 1967). Pardo referred to various geological surveys indicating that the seabed contained minerals of great value. At that time, it was not possible to mine them, but Pardo warned against a race and a parceling of the seabed into different territories. The seabed is the global heritage of mankind, he argued, and if anyone is to take advantage of it, it must be the developing countries that need this opportunity to build industry and prosperity. This last point was never incorporated into any agreement. However, the United Nations Convention on the Law of the Sea states that "The Area and its resources are the common heritage of mankind" (70).

It took more than 30 years after Pardo's speech before the ground was broken for the first extractions of minerals from the seabed. But while some people see a possible solution for a future mineral crisis, others argue for conservation, and several NGOs have asked for an international moratorium—a temporary prohibition—to control seabed mining. There are primarily three geological resources for seabed mining: manganese nodules, seafloor massive sulfides, and cobalt-rich crusts. In the High North Atlantic, there is only one of these, namely seafloor massive sulfide deposits which are primarily found at hydrothermal vents alongside the ocean ridges (Miller et al., 2018). Around such vents, it will be possible to extract so-called "sulfide content" in addition to some valuable metals (Miller et al., 2018). Most hydrothermal vents are located on various states' continental shelves. When the Norwegian continental shelf was expanded, several of these deep-sea vents were incorporated into the Norwegian territory.

Media and Territory

Territory and mediation are intertwined in a way that makes it difficult to grasp the concept of territory detached from mediation.

As several human geographers have stated, territory is not a fixed unity. It is mutable and fluid (Elden, 2013, p. 330) and "always becoming" (Painter, 2010, p. 1094). According to the geographer Stuart Elden (2010), two factors are necessary to maintain a territory: a developed legal framework, and different techniques or technologies, especially monitoring and mapping techniques. Elden argues that this is especially important in cases where there are inhabited or inaccessible areas that need to be built under a state's control, and he uses tundra, desert, and mountainous areas as examples (Elden, 2010, p. 809). We can also add the seabed to that list.

While Elden is most concerned with the political or governmental aspects of territory, geographers such as Antonsich (2010, 2017), Brighenti (2010), and Raffestin (1984, 2012) underline that territory (also) has to be understood as cultural identity. Territory is the result of a group of people having a common identity attached to an area, while at the same time strengthening and maintaining that identity. As Brighenti (2010, p. 66) states: As an act of performance, territory requires interaction, actors and often an audience. Using Anderson's (1991) concept of "imagined communities," he argues that territory is "an imagined (not imaginary) entity" (Brighenti, 2010, p. 58). According to this view, territory is a conception shared by a group of people. An understanding of territory as an imagined and shared entity can be a useful tool to describe the role of mediation in the performing of territory (Alnæs, 2021). To maintain a territory, not only a legal framework and mapping and surveillance techniques are needed, but it is also necessary with stories and imaginations about what the territory is, and thus who it belongs to. For Brighenti and Kärrholm, the visual dimension is of particular importance "territories have to do with displaying, presenting, and becoming visible to someone for some purposes or in some respects or capacities" (Brighenti and Kärrholm, 2020, p. 31).

But what about the places which are hard to depict, or from which there are sparse stories to tell? In an early discussion of media's role in the production of territory, the Swiss geographer Claude Raffestin launches a concept of mediators. Territiory, Raffestin argues, is a result of the way an individual or a group of people relates to the surroundings. It is an assembly of relations to the concrete space or the social organized space. Thus, territory depends on what Raffestin names *mediators*, which are all forms of connection between the individual and the surrounding's technical instruments and all kinds of representations. "Territory can be considered as space informed by the semiosphere—that is, the signifying system from which the actor draws the informational resources for action" (Raffestin, 2012). With a twist on Wittgenstein's statement, "The limits of my language mean the limits of my world" (2021, p. 204), Raffestin claims: "The limits of my territoriality are the limits of my mediators" (Raffestin 1984, p. 141).

When geological formations or biological material on the seabed are the theme of a music performance, the territory is expanded by our mediators, to use Raffestin's concept. It is not the geological formation that becomes music. The sound refers to an assemblage that includes the geological formation, but also the cultural understanding of it produced through robots, digital images, news articles, and music composition, through research reports, white papers, and maps. The part of the seabed is created and acquired through various cultural measures, through mediation and administration.

In Raffestin's conception, "mediators" refers to all entities or factors that facilitates, influences, or modifies the interactions and relationships between humans and their environment. This could include sensing technologies and technical tools, social structures, symbolic frameworks and text—be they verbal, musical or visual—that help shape how individuals and groups perceive, access, and manage space and resources. Raffestin's idea underscores the importance of these mediators in understanding how territorial relations are constructed and transformed.

Case 1: Appropriating the Seabed

The incorporation of the extended continental shelf in 2009 did not get much attention in Norwegian news media. Some news articles were published, but the lack of images and stories made the texts short and focused on the juridical aspects of the extension. A few more articles were produced when some images from the newly appropriated seabed were made accessible in subsequent years. "Here comes 'New Norway'" stated the newspaper Dagbladet in a headline (August 2, 2013). The statement had a double meaning: parts of the volcanic seabed being new in a geological sense, and it being newly incorporated into Norway. The *newness* was a recurring point in several publications (Wehus, 2013; Dahlen 2021, p. 147).

Still, it was not before a wider specter of videos, pictures, accounts, and scientific observations were produced that the extended continental shelf really became a new part of Norwegian territory in a Raffestinian meaning.

In May 2021, one lengthy journalistic article describing the deep-sea vents was published in two Norwegian regional newspapers; Bergens Tidende and Stavanger Aftenblad.[2] Both of the newspapers are located on the coast of Western Norway in the cities of Bergen and Stavanger, which are two Norwegian gateways to the High North Atlantic. They are the two most widely-read regional newspapers in this part of the country. The actual seabed had previously been described in science communication websites and in shorter newspaper articles, but this was the first major coverage in regional or national newspapers.

The article was available online and in print in the Saturday magazines. The texts in the four publications are more or less the same, but the online articles also include videos and links to other articles and other webpages. Thus, the online articles are a kind of immersive journalism (Gynnild et al., 2021) allowing the reader to dive into the topic through reading, scrolling, clicking, and watching videos.

All four publications used the same title: "This could become Norway's new treasure chest. The battle over deep-sea mining has now begun."[3] The newness is still a central aspect in the framing of the area, but this time as a new resource and new economy.

In Bergens Tidende's online publication, the title is in white letters on a mainly dark blue picture of the seabed (Andersson et al., 2021). The picture covers the whole screen (see Figure 10.2).

The picture has similarities with photos of shipwrecks which have been in the sea for years. The clay-grey, almost sculptural form stretching upwards and the dark background with white particles illuminated by artificial light makes it clear that this is an underwater picture. But that is all an untrained eye can get out of it. What the motive is, where it is, and how deep it is are difficult to ascertain. The only function of the picture is the underwater link, but together with the words from the title: *mining, treasure*, and *Norway's*, the setting is established.

Without further introduction, the main text starts describing one specific site on the seabed:

Figure 10.2 Heading and first picture from the article in Bergns Tidende, May 26, 2021.
Note: Translated from Norwegian: "could become Norway's new treasure chest. The battle over deep-sea mining has now begun".

> Valuable metals are found around volcanic chimneys and in crusts on the seabed.
>
> Gold, silver and copper.
>
> But cobalt is also found here, as well as a lot of other rare metals that the world needs—for solar cells, electric cars and wind turbines.[4]

Placing gold, silver, and copper on a single line gives important signals: This is not news journalism, but more like a feature article, it will elaborate poetic techniques to create moods and evoke feelings rather than arguing in the traditional way with complete sentences. Even though gold and silver are found at Loki's Castle, the three economically most important minerals are copper pyrite, isocubanite, and zinc blende (Oljedirektoratet 2021; Snook et al., 2018). Thus, by underlining the three metals gold, silver, and copper, the journalist makes the discoveries adventurous and mythic. The three metals are often named the three coinage metals, and are a trinity known from Norwegian folklore, often representing three levels of power or evilness in the troll's world.[5] Therefore, the catalogue of the three metals may both underline the Norwegian folklore aspect of this story and connect it to the hunt for hidden treasures from the past.

Nevertheless, the explicit reason for mentioning these metals is that they are among the resources needed for the shift towards a green economy.

> But there is also cobalt here, as well as a bunch of other rare metals that the world needs—for solar cells, electric cars and wind turbines.[6]

The article continues to draw attention to Norwegian folklore and treasure hunting.

> There are hot springs here that look like chimneys. They spew out mineral-rich water that has been heated in the Earth's volcanic interior.
>
> The special natural phenomena have been given names such as Loki's Castle, Soria Moria and Ægir's Source.[7]

The term "such as" implies that the three names have something in common. Loki is a central god in Norse mythology, known for being rather unreliable, sometimes a friend, sometimes malicious. Ægir, or Æge, is the god of the sea. While Loki and Ægir are figures from the pre-Christian Norse mythology, Soria Moria derives from a Norwegian folk tale. Soria Moria Castle is a name of a troll's castle in the tale, but also the name of the tale itself, and maybe even more widely known a series of paintings by Theodor Kittelsen titled "Til Soria Moria Slott" (from the year 1900). *Soria Moria* has been an established trope in Norwegian, used by authors like Henrik Ibsen and Knut Hamsun as an imaginative place of wealth and fortune.[8] Besides, Soria Moria labels a concert venue in the center of Oslo and a well-known conference center on the outskirts of the city. The name of the conference center is the origin of the so-called Soria Moria

Declaration. This was a well-known political agreement negotiated by three government parties under the leadership of Jens Stoltenberg in 2005. The agreement had several political consequences, but was also criticized for its unrealistic ambitions, including the plan to abolish poverty. Together with the folk tale, the Soria Moria Declaration has contributed to Soria Moria having connotations of both something utopian and something quintessentially Norwegian.

Common to the place-names *Loki's Castle, Soria Moria* and *Ægir's Source* is not their origin or connection to a specific religion, but the fact that they all, among other things, refer to something ancient and culturally Norwegian (see Meyer and Lindström's chapter in this volume). Even though there are several names in the area retrieved from Norwegian folklore, there are also quite different names for underwater formations, such as *Squid Forest* or *Copper Hill*. Those names are not mentioned in the article. Such way of picking the details that strengthen the Norwegian connection clears the ground as it simultaneously constructs a Norwegian place (Alnæs, 2017). In this way, the "new Norway" is implemented in the old, everlasting, and mythical Norway.

One of the article's pictures of the seabed shows plantlike "sea feathers" and is followed by this caption:

> The seabed is also biologically rich. These sea feathers are around two meters high. There are also mounds of metal-containing minerals here, which in shape and size are reminiscent of Viking Age burial mounds.[9]

Why compare mounds on the seabed with Viking Age burial mounds? In shape and size, there are of course some similarities, but there are also some common features between the history of burial mounds and the possible future of the seabed's mineral mounds. Both contain or have contained treasures. And while a lot of plundered burial mounds are left as memorials of greed, the discussion now stands as to whether the mineral mounds should end up in the same way. Comparing the mineral mounds with Viking burial mounds will maybe make this connection more apparent for the Norwegian readers. There are burial mounds impacted by looting several places on the Norwegian mainland. Thus, the text implies the question: should we plunder this part of our country as well? The small link to Viking burial mounds and not just burial mounds shifts the focus from global responsibility to national responsibility, from *our heritage* in the sense of the world's heritage to *our heritage* in the sense of Norwegian heritage.

To turn the seabed into "the new Norway" is not only a result of the juridical expansion of the continental shelf. To become part of the cultural territory of Norway, the seabed must be imprinted with Norwegian identity. Thus, through texts like the one discussed here, the seabed takes its first steps towards what the geographer Bernard Debarbieux names a *place of condensation*. Such places feed the reproduction and stability of the nation, and are practiced "in images through books, newspapers, photos and televised documentaries, sometimes with a comparable emotional entity" (2019, p. 114). And even if some steps remain before the expanded continental shelf has such a symbolic effect as Debarbieuix's places of condensation when the news article was published, this part of the seabed already participated in reproduction of the nation.

As the geomedia scholars Paul C. Adams and André Jansson have pointed out, a place can also be understood as a medium: "If we understand 'media' to be much more than simply technologies, then places are media in their own right, since they gather, include, omit and re-collect elements of collective memory" (Adams and Jansson, 2012, pp. 308–309). While national symbols, myths, and perspectives are reflected in the seabed, the seabed has a function as a strengthening of the national imaginary. Consequently, the seabed itself can be a medium in the sense Adams and Janson point out.

Case 2: Mediating the Deep Sea

"The battle of deep sea mining" described in the article discussed above accelerated in Norway during 2023. In a white paper published in June 2023, the Norwegian government proposed to allow deep-sea mining (Meld. St. 25 (2022–2023)). Not surprisingly, the proposition resulted in news articles and several opinion pieces during the summer and autumn. Public debates were arranged in the cities of Oslo, Bergen, Stavanger, and Arendal. However, in order to open the Norwegian seabed for mineral extraction, the government needed support from the parliament, and most of the political parties had still not made a decision. The battle concerning deep-sea mining was still ongoing.

In the early autumn of 2023, two lengthy feature articles let readers dive into the actual depths of the sea. One of them was published in the left-wing newspaper *Klassekampen* and one by the public service channel *Norwegian Broadcasting Corporation* (NRK). Both offered a close encounter with the deep sea, partly similar to the one we met in *Bergens Tidende*. The articles were so-called longreads and included several pictures from the seabed. While the article in *Klassekampen* was published both in print and online, the NRK article was only published online, and could draw on the benefits of scrollytelling (Seyser and Zeiller, 2018), including overlapping text, pictures, and videos. Both articles emphasized the worries about the consequences and used the journey from the surface of the water to the seabed as a structuring plot. But while the article in *Klassekampen* focused on the conflict between financial gain and ecological demolition ("Gold rush in deep water"), NRK told the story from the perspective of the robot which was sent to map out and register the geological and biological features of the area. Due to the fact that NRK is one of the most widely-read Norwegian online newspapers and the scrollytelling format enables new encounters with the seabed, this chapter now turns to a close reading of the NRK article "In deep water" (Knežević and Helledal, 2023).

The title "In deep water" (In Norwegian: "På dypt vann") refers to the literal depths of the sea, but also refers to being in trouble. Simultaneously, there are connotations of the idiom "to be in over your head" (in Norwegian "ta seg vann over hodet" literally means to "take water over your head"), which refers to difficult situations we may have placed ourselves into. Against the background of the ongoing debate on deep-sea mining, which increased during the preceding months, it is reasonable to understand the precarious situation regarding the potential extraction of the seabed's resources. Thus, the title signals a more critical approach to deep-sea mining than the one discussed above. Nevertheless, it is striking how the point of departure in the NRK article is contrary to the articulated intention of the music "Loki's Castle." While the composer wanted to underline the connections between humans and the species on the seabed, the NRK article underlines the *unfamiliar* aspect of the same. On a video background of moving water in a grey and purple landscape, the text opens:

> It looks like a portal to another dimension
> The robot Ægir, with its red eyes, is on a journey of discovery
> What it finds can have major consequences.[10]

The sci-fi references are prevalent throughout the article. The main story is about two robots on missions to an alien world, and the allusions to sci-fi tales are apparent: one of the submarine robots is explicitly compared to the Pixar movie-robot WALL-E (2008). And just like WALL-E and other sci-fi robots, Ægir is a robot with a name and a personality: "When Ægir is not down in the depths, he lives at Havlab in Bergen. Here he is waiting to be sent out on a new voyage."[11] Havlab literally means "Ocean Lab," and refers to the Norwegian Ocean Observation Laboratory at the University of Bergen.

In addition, the other robot is given life and portrayed as a human with the body of a vehicle: "Like most excavators, the Seabed Excavator has belts instead of legs, an arm that can be rotated back and forth."[12]

The two robots share the same mission: scrutinize and report on the conditions on the seabed. The robot Ægir is described as a male (according to the article it is a "he") who lets us see the underwater world: "The robot helps us 'see' things that are hidden from our eyes."[13] This metaphor of the eyes derives from two reflections of red light seen in the pictures taken by the camera installed on the robot (see Figure 10.3). Seeing becomes a key metaphor used throughout the article. It anthropomorphizes the robot Ægir. Ægir "sees," it does not register, scan, or document. Thus, the lens is his eyes. Seeing is also a metaphor for investigating and mapping, well-established in Western culture (Crary, 1992). In such a way, according to the article, the robot Ægir registers first and foremost what is visible.

Figure 10.3 This picture was published as a part of the NRK online article "In deep water" ("På dypt vann")

Note: Any opinions expressed here are solely those of the authors and do not necessarily represent the views of the Centre for Deep Sea Research.

Source: CDeepSea/University of Bergen/ROV Aegir6000.

The body text opens with an allegory where seeing is manifested as a metaphor:

> Imagine being ordered to map an entire forest and everything that lives there. Then you are sent out in the middle of the night, with a flashlight in your hand.
>
> That is pretty much what it is like for humans to examine the seabed.
>
> Therefore, Ægir is needed. He is very popular; for the scientists, he is quite a catch.[14]

We see things not just through the eyes of Ægir, but also with the help of light emitting from Ægir. And in Ægir's light, things become different. Scrolling through the text above, the reader reaches a video, first of a fish, slowly moving past a grey sandhill, and then a picture of dozens of white creatures settling on a stone. A short text appears in the middle of the picture: "Countless shrimp-like creatures have settled. White, almost ghostly in the light of Ægir."

The layers of mediation and representation are becoming complex and difficult to comprehend: the light of Ægir allows us to see with Ægir's eyes, but then we see a world that is distorted, and it resembles something supernatural. Consequently, the article launches a way to understand the ocean: as exceptions, something twisted, that can resemble supernatural (see Figure 10.3).

Underlining the strangeness is also a way of understanding or a way of grasping the unfamiliarity of the deep sea. Through allusions and explicit comparisons, the account of the deep sea appears as ghost-tales and sci-fi movies. The anthropologist Stefan Helmreich (2009) has discussed how the figure of the alien has been used by marine biologists and others to make knowledge about the ocean meaningful: "The alien is a channel for exchange between the oceanic and the human" (p. xi). In other words, by acknowledging the seabed's unfamiliarity, it becomes understandable.

It is not only the emphasis on the alienness that is striking in this article, it is also the emphasis on mediation. The multimodal text does not primarily thematize what is found on the seabed, but how we can gain knowledge about it, and which role mediation plays in this knowledge acquisition.

The story of the mediation of the strange, alien deep-sea world plays out against a clear backdrop: the seabed appears to be something Norwegian as in the article in *Bergens Tidende*. The fact that this is a new part of Norway's territory—only 14 years old—is not thematized, but several times it is explicitly and implicitly stated that it belongs to Norway: "Perhaps there is a new industrial adventure for Norway in the depths—which can also help slow down climate change?"[15]

The term "industrial adventure" or in Norwegian "industrieventyr" is often used to describe how an industry or a company develops and grows rapidly. The term is almost a cliché in the genre of corporate history, where industrial adventure most often refers to industry in connection with the extraction of power, oil or minerals. As the term "eventyr" is used for both adventure and fairy tale, "industrieventyr" has connotations of something unreal, but also something well-known and recognized in Norway. The Norwegian folk tales are seen as a central part of the nation, and an industrial adventure is often perceived as something particularly Norwegian. The most famous Norwegian folk tales are those about *Askeladden*—the poor, exploratory boy who ends up overcoming the trolls and marrying the king's daughter, thematizing the growth from zero to the kingdom's most powerful. The term *industrial adventure* appears in many of the texts about seabed mining, and many of the perspectives are brought together in this term: the unbelievable, the national, and the possibilities for wealth and growth.

The title "In deep water" can of course be understood literally, as this is an article about what is going on in the deep sea. However, with the battle concerning deep-sea mining as a background, the title can also be understood as an idiom: someone is getting into deep trouble. On the other hand, it is also, as we have seen, a story about the limits of mediation. Thus, the trouble can be a question of understanding and grasping the deep sea.

The Media Territory of the Extended Norwegian Continental Shelf

Let us return to the ceremony hall in Oslo. After the performance of music by Helmut Lachenmann and Richard Strauss, the singers in The Norwegian

Solist's Choir positioned themselves along the walls around the hall so that the singers surrounded the audience. They then started performing what was announced as the evening's main attraction: the composition "Loki's Castle." It opens with breathing sounds resembling waves and wind. We then hear whining, not unlike renderings of whales. The printed program describes how some whale species can dive down to 3000 meters "and are vivid connections between humans and organisms at great depths" (The Norwegian Soloist's Choir, 2023, p. 2). Gradually, the music becomes more diverse and intense, and it can probably be understood as a representation of life at the bottom of the sea. The arrangement of the choir creates an immersive effect, taking the audience on a metaphorical dive.

The movement described by the music is the opposite of the one in the two journalistic texts: from the surface down into the depths. All those vertical journeys through the water column are equivocal: they appear as journeys through territories or territorialities, journeys from culture to nature or from center to periphery, from the ceremony hall of the University of Oslo to the deep seabed of the High North Atlantic, or from the seabed to the researchers and politicians of Norway. Those journeys through the water column connect the seabed with the mainland, the former Area with the Norwegian State. As a result, both the concert and the two articles mediate the ocean as a medium: the sea facilitates connection between the seabed and the surface, and between life on the seabed and Norwegian cultural heritage.

Several researchers have drawn parallels between how space and text are made up of earlier stories, shared memories and cultural references. The philosopher Henri Lefebvre (1991) has used the terms *networks* and *paths* to describe social space and draws parallels to how a text is a weave. "Paths are more important than the traffic they bear, because they are what endures in the form of the reticular pattern left by animals, both wild and domestic and by people" (p. 118), Lefebvre argues and asks if such patterns can be called a text or a texture. These thoughts are followed up on by the literary scholar Bertrand Westphal who uses the concept of intertextuality to show how space is composed of past experiences and bits of cultural heritage, and is full of references in the same way as texts. "Space corresponds to a texture, because it is reticular down to its smallest folds" (Westphal, 2007, p. 163). Thus, in Westphal's eyes, there is a constant oscillation between text and space: "Calvino makes the Paris that makes Calvino, Butor makes the Tokyo that makes Butor, Borges makes the Buenos Aires that makes Borges, and so on to infinity." (Westphal, 2007, p. 163). Therefore, space is both textual and physical, or "Of stone and paper" (Westphal, 2007, p. 153).

But to what extent is it relevant to distinguish between space and text? Space consists of texts about space, but also physical space or what Lefebvre calls geometric space. But the texts about space are, of course, not detached from this "real space," even though the real space cannot be grasped without references to culture and text. Geographer Edward W. Soja (1996) uses the term *thirdspace* to blur the distinction between representation and what is represented. Space is both real and imagined simultaneously, he argues.

Nevertheless, some spaces are more or less inaccessible outside the texts. This is the case with outer space, the Antarctic continent and the deep sea. If I am going to question a statement about those places, I must consult other mediations. The cities from Westphal's examples can be sought out, one can even settle there, but the seabed is not accessible to anyone other than through media. The seabed is *only* mediated. "The new Norway," the newest part of the Norwegian territory, exists as territory, only in a mediated state, or only as text; images, film, songs, news and opinion articles, and as performances derived from these. In this way, the seabed is a media territory.

During the late autumn of 2023, it became clear that a majority in the Norwegian parliament would give green light for mining to take place on the Norwegian deep seabed. The week after, the EU Parliament held a plenary debate on the Norwegian decision (European Parliament, 2024) and in the end of the month, they launched a motion for a resolution which "Expresses its concerns" regarding Norway's decision. From a territorial perspective, there are two interesting aspects in the EU resolution. First, although Norway has territorial rights, the seabed is regarded as something valuable everyone has a responsibility to take care of. Hence, there is both a sense of the juridical and cultural aspects of territory in the statement. Second, the volumetric aspect of territoriality is emphasized. "whereas … much of the area above the extended continental shelf is considered as high seas and areas for international fisheries" (European Parliament, 2024). In other words, a volume-territory perspective that has been absent from the Norwegian texts scrutinized in this chapter.

However, in December 2024, the Norwegian government was forced by a Norwegian opposition party to reverse their decision about deep-sea mining. So far, there will be no mining on the seabed of the Norwegian Sea.

It is difficult to speculate whether media coverage of this area has influenced the political decision to allow—and then deny—deep-sea mining. Nevertheless, there is little doubt that commentaries, news articles, feature articles, immersive journalism, and even mining-critical opinion pieces have helped to consolidate that this area is Norwegian and the view that Norway has the right to control this part of the seabed and can do as they please, regardless of what others may argue.

Notes

1 In Norwegian, the area is called "Smutthavet" (The Loop Sea), which is an allusion to the more contested ocean area called "Smutthullet" (The Loop Hole) in The Barents Sea.

2 Through collaboration, these newspapers regularly publish the same texts. The cooperation also includes other regional and national newspapers.

3 In Norwegian: "Dette kan bli Norges nye skattkammer. Nå begynner slaget om gruvedrift på dypet."

4 In Norwegian "Rundt vulkanske skorsteiner og i skorper på havbunnen ligger det verdifulle metaller. Gull, sølv og kobber. Men her finnes også kobolt, samt en drøss andre sjeldne metaller som verden trenger – til solceller, elbiler og vindturbiner."

5 See: "Kari Trestakk" (Asbjørnsen and Moe, 2009, p. 163), "Jomfrua på glassberget" (Asbjørnsen and Moe, 2009, p. 246) and "Mestermø" (Asbjørnsen and Moe, 2009, p. 274).

6 In Norwegian: "Men her finnes også kobolt, samt en drøss andre sjeldne metaller som verden trenger—til solceller, elbiler og vindturbiner."

7 In Norwegian: "Her er varme kilder som likner skorsteiner. De spyr ut mineralrikt vann som er varmet opp i jordens vulkanske indre. // De spesielle naturfenomenene har fått navn som Lokeslottet, Soria Moria og Ægirs kilde."

8 See Henrik Ibsen: *Peer Gynt* and Knut Hamsun: *Pan*.

9 In Norwegian: "Havbunnen har også biologisk rikdom. Disse sjøfjærene er rundt to rundt meter høye. Her finnes også hauger med metallholdige mineraler som i form og størrelse minner om vikingtidens gravhauger."

10 In Norwegian: "Det ser ut som en portal til en annen dimensjon Roboten Ægir, med sine røde øyne, er på en oppdagelsesferd. Det den finner kan få store følger."

11 In Norwegian: "Når Ægir ikke er nede i dypet, bor han på Havlab i Bergen. Her venter han på å bli sendt ut på nytt tokt."

12 In Norwegian: "Som gravemaskiner flest har Seabed Exacavator belter som bein, en arm som kan roteres hit og dit."

13 In Norwegian: "Roboten hjelper oss å 'se' ting som er skjult for våre øyne."

14 In Norwegian: "Se for deg at du blir beordret til å kartlegge en hel skog og alt som lever der. Så blir du sendt ut midt på natta, med en lommelykt i hånda. Omtrent sånn er det for mennesker å undersøke havbunnen. Derfor trengs Ægir. Han er en populær type, for forskerne er han en skikkelig catch."

15 In Norwegian: "Kanskje finnes det et nytt industrieventyr for Norge i dypet—som attpåtil kan bidra til å bremse klimaendringene?"

References

Aarnes, H. (2023). Hydrotermiske undervannskilder. *Store Norske Leksikon*. https://snl.no/hydrotermiske_undervannskilder.

Adams, P.C. and Jansson, A. (2012). Communication geography: A bridge between disciplines. *Communication Theory*, 22(3), 299–318. doi:10.1111/j.1468-2885.2012.01406.x.

Alnæs, J. (2021). *Tekst og territorium. Det norske Antarkis i norske nyhetstekster 2005–2019* [PhD thesis] University of Oslo.

Alnæs, J. (2017). The prime minister of Antarctica: Entrenching territory through journalistic travelogues. *Studies in Travel Writing*, 4(2017).

Anderson, B. (1991). *Imagined Communities. Reflections on the origin and spread of nationalism*. Verso.

Andersson, A. (2022, July 3). Macron vil stanse gruvedrift i havene—Vi trenger mineralene, sier Støre. *Bergens Tidende*. https://www.bt.no/nyheter/okonomi/i/Qy1kpx/macron-vil-stanse-gruvedrift-i-havene-vi-trenger-mineralene-sier-stoere.

Andersson, A., Alvheim, H. Vegum, Lambrechts, L., and Bøe, B. (2021, May 26). Dette kan bli Norges nye skattkammer. Nå begynner slaget om gruvedrift på dypet. *Bergens Tidende*. https://www.bt.no/btmagasinet/i/R97Jyx/dette-kan-bli-norges-nye-skattkammer-naa-begynner-slaget-om-gruvedrift-paa-dypet.

Antonsich, M. (2010). Rethinking territory. *Progress in Human Geography* 35(1), 422–425. doi:10.1111/j.1467-8330.2010.00795.x.

Antonsich, M. (2017). Territory and territoriality. In D. Richardson, N. Castree, M.F. Goodchild, A. Kobayashi, W. Liu, andR.A. Marston (Eds.), *The international encyclopedia of geography.* doi:10.1002/9781118786352.wbieg0022.

Asbjørnsen, P. Chr. and Moe, J. (2009). *Samlede eventyr: Andre bind.* Gyldendal.

Ashcroft, B. (2005). Forcing newness into the world: Language, place and nature. *ARIEL: A Review of International English Literature*, 36, 1–2.

Adresseavisen (01.08.2013). Har funnet Norges Yellowstone på havbunnenhttps://www.adressa.no/nyheter/innenriks/i/346g3M/har-funnet-norges-yellowstone-pa-havbunnen.

Aftenposten (01.08.2013). Har funnet unikt vulkanområde med metaller verdt milliarderhttps://www.aftenposten.no/norge/i/212qB/har-funnet-unikt-vulkanomraade-med-metaller-verdt-milliarder.

Brighenti, A.M. (2010). On territorology: Towards a general science of territory. *Theory, Culture and Society*, 27(1), 52–72.

Brighenti, A. and Kärrholm, M. (2020). *Animated lands: Studies in territoriology.* University of Nebraska Press.

Crary, J. (1992). *Techniques of the observer: On vision and modernity in the nineteenth century.* MIT Press.

Dahlen, P.V. (2021). *Verden under vann: En oppdagelsesreise i havet.* Vigmostad Bjørke.

Debarbieux, B. (2019). *Social imaginaries of space: Concepts and cases.* Edward Elgar Publishing.

Elden, S. (2010). Land, terrain, territory. *Progress in Human Geography* 34(6), 799–817. doi:10.177/0309132510362603.

Elden, S. (2013). *The birth of territory.* University of Chicago Press.

European Parliament (2024). *Motion for a resolution on Norway's recent decision to advance seabed mining in the Arctic.* (2024/2520(RSP)) https://www.europarl.europa.eu/doceo/document/B-9-2024-0095_EN.html.

Fransson, L. (2022, November 13). Macron vil stoppe norsk industrieventyr. *Børsen. Dagbladet.* https://borsen.dagbladet.no/nyheter/macron-vil-stoppe-norsk-industrieventyr-1/77690907.

Gynnild, A., Uskali, T., Jones, S., and Sirkkunen, E. (2021). What is immersive journalism? In A. Gynnild, T. Uskali, S. Jones, and E. Sirkkunen (Eds.), *Immersive journalism as storytelling: Ethics, production, and design.* Routledge.

Helleland, B. (2000). Place-names (geographical names) on and off the COAST OF WESTERN Norway. Names of oil fields compared to traditional place-names. Living with Diversity. 29th International Geographical Congress. 14–18 August 2000. https://docslib.org/doc/2692632/on-and-off-the-coast-of-western-norway-names-of-oil-fields-compared-to-traditional-place-names.

Helmreich, S. (2009). *Alien ocean: Anthropological voyages in microbial seas.* University of California Press.

Knežević, M. and Helledal, E.J. (2023, August 26). På dypt vann. *NRK.* https://www.nrk.no/klima/xl/ned-i-dypet-1.16472301#intro-authors–expand.

Lefebvre, H. (1991) [1974]. *The production of space.* Blackwell.

Meld. St. 20 (2019–2020). *Helhetlig forvaltningsplaner for de norske havområdene.* Barentshavet og havområdene utenfor Lofoten, Norskehavet, og Nordsjøen og Skagerak. https://www.regjeringen.no/no/dokumenter/meld.-st.-20-20192020/id2699370/.

Meld. St. 25 (2022–2023). *Mineralverksemd på Norsk kontinentalsokkel – opning av areal og strategi for forvaltning av ressursane.* https://www.regjeringen.no/no/dokumenter/meld.-st.-25-20222023/id2985856/.

Miller, K.A., Thompson, K.F, Johnstone P.S., and Santillo, D. (2018). An overview of seabed mining including the current state of development, environmental impacts, and knowledge gaps. *Frontiers in Marine Science*, 4(2018). doi:10.3389/fmars.2017.00418.

Oljedirektoratet (2021). Teknologirapport havbunnsmineraler. https://www.regjeringen.no/contentassets/a3dd0ce426a14e25abd8b55154f34f20/teknologirapport-havbunnsmineraler_dnv.pdf.

Painter, J. (2010). Rethinking territory. *Antipode*, 42(5), 1090–1118.

Pardo, A. (1967). Speech before the UN General Assembly. UN General Assembly, First Committee.

Raffestin, C. (2012). Space, territory and territoriality. *Environment and Planning D: Society*, 30(1), 121–141.

Raffestin, C. (1984). Territoriality: A reflection of the discrepancies between the organization of space and individual liberty. *Freedom and Boundaries*, 5(2), 139–146.

Seyser, D. and Zeiller, M. (2018). Scrollytelling—An analysis of visual storytelling in online journalism. 22nd International Conference Information Visualisation (IV), Fisciano, Italy, 2018, pp. 401–406, doi:10.1109/iV.2018.00075.

Snook, B., Drivenes, K., Rollinson, G., and Aasly, K. (2018). Characterisation of mineralised material from the Loki's Castle hydrothermal vent on the Mohn's Ridge. *Minerals*, 8, 576.

Soja, E.W. (1996). *Thirdspace: Journeys to Los Angeles and other real-and-imagined places.* Blackwell Publishing.

The Norwegian Soloists' Choir [Det Norske Solistkor]. Program for The Norwegian Soloists' Choir's *Et hav av stemmer* at University Aula, Oslo, February 10, 2023.

United Nations (1982). *United Nations Convention on the Law of the Sea.* Article 76. https://www.un.org/depts/los/convention_agreements/texts/unclos/unclos_e.pdf.

Valo, M. (2022, July 1). Macron takes stance against deep-sea mining at UN Ocean Conference. *Le Monde.* https://www.lemonde.fr/en/environment/article/2022/07/01/macron-takes-stance-against-deep-sea-mining-at-un-ocean-conference_5988635_114.html.

Wehus, W.N. (2013).Norge kan få ny nasjonalpark under vann. *Forskning.no* (03.08.2013). https://www.forskning.no/naturvern-partner-universitetet-i-bergen/norge-kan-fa-ny-nasjonalpark-under-vann/617997.

Westphal, B. (2007). *Geocriticism: Real and fictional spaces.* Palgrave Macmillan.

Wittgenstein, L. (2021 [1921]). *Tractatus logico-philosophicus.* Anthem Press.

11 Linking to Sedna

Infrastructural Networks in Nunavut

Nicholas R. Silcox

Introduction

In the seas of the High North Atlantic, infrastructure is more critical than ever, propelling a new constellation of communications and environmental technologies. The prevalence of rural communities, the central presence of the ocean to many communities across the region, and proximity to the Arctic situate telecommunications and environmental monitoring infrastructures as uniquely necessary. Science Monitoring and Reliable Telecommunications (SMART) subsea cables bring telecommunications and environmental monitoring concerns together in a single technological infrastructure. They represent the possibility of ongoing, in-situ measurements from the deep ocean (Howe et al., 2022). In the Labrador Sea off the coast of the Northern Canadian territory Nunavut, a cable with SMART capacity is set to be installed in 2025, connecting Nunavut to Labrador and through to Norway via the Leif Erikson system (Submarine Cable Networks, 2021). The system is a project of CanArctic Inuit Networks (CAIN), a majority-Inuit controlled company started in 2021 in order to bring the cable system to fruition, including a cabled connection to Nunavut. The cable system is called SednaLink, named for the Inuit figure Sedna.

SednaLink could represent an important demonstration of a telecommunications system and infrastructural network that complicates any simple alignment with imperial, colonialist extraction—the historical and contemporary context for many infrastructural networks. I turn to SednaLink as an example of an infrastructure that might demonstrate an alternative narrative regarding infrastructure more broadly, as it brings together ocean science and indigenous actors. In what follows, I consider SednaLink in relation to its specific political economy, cultural and environmental contexts, including the lack of rural connectivity in Northern Canada and the significance of a network operated by members of the indigenous community. Doing so demonstrates its future potential as an infrastructure that addresses infrastructural needs while bringing together science and indigenous culture. Furthermore, I explore the system's specific relation to its namesake and the ways the system is situated to build on the tradition of Sedna. Her story

DOI: 10.4324/9781003478805-11

ultimately represents as a kind of infrastructure for the Inuit people that represents environmental connection and the accumulation of critical resources, such as food, energy, and transportation. I make this case through an analysis of discourse surrounding SMART cables to consider their emergence broadly, with attention to oceanographers' advocacy for SMART cables. I turn to the specific circumstances, context, and discourse surrounding CAIN and the SednaLink system, including interview responses from a CAIN co-founder, Madeline Redfern.[1] Additionally, I engage with the story of Sedna and her role in the Inuit community across various circulations of the tradition, including by CAIN, and the significance of naming the network after a figure from an oral tradition, particularly as it relates to network sovereignty and alternative infrastructural forms.

To clarify, SednaLink is still a speculative network and is not yet officially operational at the time of writing. The speculative possibility remains significant for considering the ways it points to the potential of infrastructural networks to be reimagined through wedding indigenous cultural traditions to emerging technologies and contemporary science. This chapter intends to think with the possibilities SednaLink represents, while grappling with its current realities, complications, and ideological tensions.

It is important to note that I approach the story of Sedna and the culture of the Inuit people generally as an outsider. As a *qallunaaq*, or a non-Inuit White person of European descent, writing from the stolen lands of the Lenape people, I am very much embedded within institutions and the legacies of colonialism that have occluded and marginalized the indigenous people of North America. I am not an expert in the culture of the Inuit people nor I am a resident of Nunavut or Canada. This chapter is not intended to be an authoritative discussion of Inuit culture. Instead, as a scholar of media and infrastructure, my focus is the system itself. To this end, I aim to highlight the significance of the infrastructural system in question, amplify its surrounding narrative, and point to the broader potentiality of site and culturally specific approaches to infrastructure. I believe that doing so addresses the intertwined phenomenon of digital connectivity and environmental concerns.

SMART Cables, CanArctic Inuit Networks, and Their Political Economy

SMART cables represent the most realized possibility of consistent, ongoing access to the deep ocean of any previous ocean monitoring strategy (Howe et al., 2022). They serve as the primary infrastructure for global telecommunications and traffic over 99 percent of the global internet, modified with environmental sensors installed on signal repeaters at regular intervals along the cable body to monitor environmental conditions in the deep ocean (Howe et al., 2022; Starosielski, 2015). These cables monitor temperature, pressure, salinity, vibration, acceleration, and geologic movement (Howe et al., 2022). The sensors collect data about the oceanic environment that is then transmitted along the cable to the surface, eventually to a regional monitoring station where the information is processed and interpreted. The response to this data sets in motion any number of human and nonhuman technical, social, and political practices and actions, ranging from immediate emergency services to the development of policy (Gabrys, 2015). Significantly, SMART cables are imagined by some to be the future of the global subsea telecommunications network. The very first SMART cable to be connected to the global subsea network is currently being installed and will launch in 2025, with a few others like SednaLink set to be installed or currently in planning stages. Where cable sensing is used in a number of regional contexts, SMART cables advocates, mostly oceanographers and seismologists, propose for SMART cables to eventually replace the cables that currently comprise the global network and produce a "massive earth sized sensor" (Starosielski, 2021, p. 164). Ocean scientists describe the sensors as "piggy-backing" on the global network (Howe et al., 2019). The sensors would collect information about the deep sea environment which would benefit science, but also improve the reliability of the telecommunications network by sensing potential environmental threats to a cable body. SMART cables entangle the agential, material capacities of the deep with global connectivity, blending oceanic and terrestrial signal exchange.

SMART cables have emerged from the growing trend in the past several decades to deploy subsea cables, often retired telecommunications cables, as regional oceanic monitoring systems. In these cases, a single cable is used as both the power source and the site of sensing for oceanography work. These projects are not enmeshed in the broader global telecommunications network.[3] In the past decade, oceanographers and seismologists have begun exploring integrating sensing cables into the broader global telecommunications network (Butler, 2012). These same scientists have also taken up advocacy efforts, attempting to

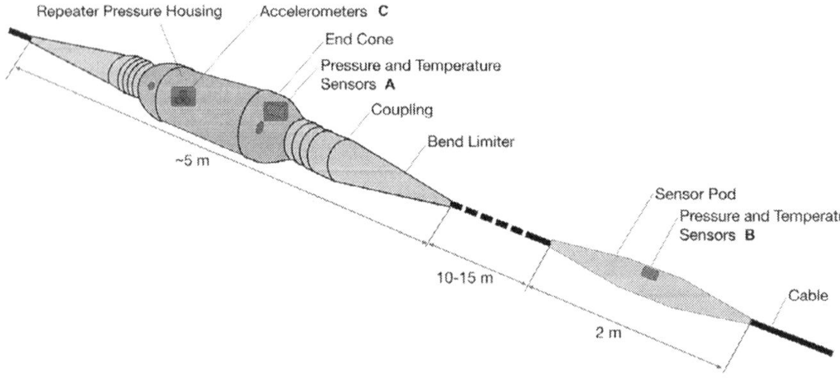

Figure 11.1 An image of the proposed SMART Cable prototypes with components labeled[2]

get buy-in from members of the subsea telecommunications industry and regional governments concerned with addressing some environmental risks, including earthquakes and climate change (Howe, 2022).[4] Advocates have also targeted areas that traditionally lack connectivity to pair environmental monitoring with expanded connectivity. The locations where SMART systems have been or will be installed are variable, but the common motivations for interest in SMART cables are environmental hazard concerns, with primarily partnerships developed between advocates, scientists, and regional governments (Howe, 2022).

SednaLink is proposed to connect the Nunavut region to the broader network for the first time, an area that lacks internet infrastructure and efficient connectivity (CRTC, 2020; Wright, 2023). Many of the rural areas in Northern Canada, including Nunavut, are indigenous territories, and connectivity issues and lack of infrastructure are vestiges of Canada's legacy of settler colonialism. This installation will be owned and operated by CanArctic Inuit Networks (CAIN), an independent, private, majority-Inuit-owned company that has set out to bring cabled internet to Nunavut to expand access to digital connectivity (Arctic360, 2022; Wright, 2023).[5] SednaLink is CAIN's first and only project — in fact, it represents their origin and motivation to incorporate in the first place. While the specific governing structure of CAIN seems to be no different than another company, their status as an independent company that is also owned by individuals with ties to the community they serve is exceptional for the subsea industry, which is largely a global industry. Likewise, many companies operate under the direction of yet larger companies or conglomerates.

CAIN's interest in a SMART system is unique in the context of the broader subsea cable industry and speaks to their specific commitment to the affordances of the infrastructure. One might assume that SMART cables would be primarily supported by industry interested in profiting from yet another form of surveillance capitalism (Zuboff, 2015) or that they represent an extension of the broader "SMART" mandate (Halpern and Mitchell, 2023); however, this is not entirely the case with SMART cables. The subsea cable industry has largely been reluctant to embrace SMART cables because of concerns regarding the capacity and logistics of data sharing with scientists and local governments, their expense, and the overall general conservatism of the industry (Howe, 2022; Howe et al., 2022). The industry has historically prioritized ongoing security and connectivity rather than operationalizing cables for other concerns. Nicole Starosielski has documented these priorities and the conservatism of the industry through the narrative strategies of "insulation" and "interconnection" they have historically deployed to ensure both security and connectivity (Starosielski, 2015). SMART cables reimagine subsea cables as actors and contribute to the production and circulation of data concerning the local environment. In "opening up" cables to the dynamic materiality of the environment around them, SMART cables run counter to the long-held emphasis on insulation within the industry. Instead, SMART cables have almost exclusively emerged from partnerships between ocean and geological scientists and international civic institutions like the UN, as well as regional governments and institutions rather than from private industry. Scientists and public actors have sought to secure funding in the form of grants, support from organizations like the World Bank, and governmental funds, while continuing to publish articles that advocate for the potential benefits of a SMART global network (Howe, 2022).

CAIN's motivation for adopting SMART cable technology and their approach to funding the SednaLink project reflects the interests of the scientist-advocates who have most strongly supported SMART cables. Most urgent for Nunavut and CAIN is the lack of cabled connectivity in the region. The nonexistent or ineffective internet and telecommunications infrastructure in the region forecloses the social and economic opportunities and advantages the internet facilitates (Haight et al., 2016). Prior to 2023, Nunavut was the only jurisdiction in Canada that did not have access to internet speeds over 25 megabits per second (Lochead, 2023). Now, some individuals have access to faster speeds via individualized satellite access, but most still rely on shared satellite traffic that can reach up to 15 MPBS (Tranter, 2023; Arctic360, 2022). While there have been efforts to address the connectivity gap via satellite infrastructure, satellite connectivity is inefficient, ineffective, and more expensive for individual consumers than cabled connection (Flynn, 2021; Lochead, 2023). StarLink, the satellite connectivity program and subsidiary of SpaceX, is the primary provider of satellite connectivity, and it is expanding in the region. The expense and tiered capacity pricing of StarLink reifies the very inequities that are the original source of ineffective access (Lochead, 2022b). Nunavut is also the only Canadian territory that relies exclusively on satellite connectivity (NTI, 2020). A potential second cable to connect Nunavut, the Katittuq Nunavut Fibre (KNF) project, was recently placed on indefinite hold (Tranter, 2023). The KNF project was intended to be publicly funded and operated by the Nunavut government. However, the expenses exceeded the proposed budget and the Canadian government decided to halt the project, continuing a long legacy of disinvestment by the Canadian federal government in the infrastructure needs of Nunavut.

In addition to providing efficient and effective connectivity to the region, CAIN has been explicit about their interest in the environmental benefits afforded by a SMART system. In particular, they have highlighted the benefits of monitoring the seas around Nunavut to assess the impact of climate crisis on the ocean, to track and monitor whale migrations as they change, to protect against potential environmental harms, and to ensure security in the region in collaboration with local government and scientists (Arctic360, 2022). Following the twin pieces of legislation known as the Nunavut Land Claims Agreement and the Nunavut Act of 1993, the Inuit people govern the territory and have been given federally recognized sovereignty over issues regarding land usage and environmental concerns in the territory.[6] CAIN is among several local actors that argue for the urgent need for functioning infrastructure to serve Nunavut, which has experienced limited and failing infrastructure in many critical areas, leaving the territory unable to address the unique challenges of contemporary life in the Arctic.[7] The dual capacity of SMART cable connection produces two infrastructures in a single technical system: SednaLink would address connectivity issues in the region, and its SMART capabilities could potentially empower the territory as an active participant in addressing climate change—or other environmental or security issues through environmental monitoring and data collection.

Developing any kind of cabled connection requires significant planning, development, and building costs. These concerns are major obstacles for any cable, but particularly so for a region like Nunavut that has historically been neglected. SednaLink is no exception to these concerns. Its construction has experienced some issues with financing. CAIN has and continues to collect investments from the likes of government funding, nonprofit grants, and private equity, including seed organizations like the Ocean Startup Project and even potentially an indigenous development corporation (Bell, 2021; Lochead, 2022a; Crunchbase, 2023). As of 2024 at the time of this writing, the project is still delayed but appears to be continuing to move forward following a recent partnership between CAIN and an Alaskan company specializing in fiber optic networks (Wright, 2023). A standard subsea cable would be easier and more affordable to install, particularly for a location that lacks any cabled connection and a new company without other assets or large sums of startup capital compared to the tech giants, including Google, Facebook, Microsoft, and Amazon, that increasingly comprise a large percentage of subsea infrastructure development (Telegeography, 2021). Given these obstacles, the decision to pursue a SMART system that is even more logistically complicated and expensive seems exceptional and more complicated than simple profit motive.

Part of CAIN's interest in SMART cables is rooted in concerns around indigenous sovereignty and culture. CAIN has described the benefit of data collection as a way for Nunavut to have more economic capacity in the territory and to have a more defined role in global politics, especially where it concerns the Northern Atlantic (Arctic360, 2022). Where this is possible, it is through their partnerships with government, science, and the various institutional actors in Nunavut. Notably, this is distinct from the way large tech companies approach infrastructural development where they act more independently. There is reason to be skeptical of CAIN's appeal to economic sovereignty as the company is not democratically owned by the broader Nunavut community and therefore is not a direct reflection of community self-determination. CAIN's self-described interest in security appears to affirm the contemporary geopolitical landscape and imagines SednaLink as positioning Nunavut as a geopolitical actor. While geopolitical security is a legitimate concern, this interest speaks to the limitations of infrastructure to produce alternative political forms and the ongoing, fraught relationship between science and geopolitics. Despite this, an Inuit-owned company in collaboration with an Inuit majority government appears to be a significant step toward something like self-determination than infrastructure owned and operated by external actors like tech companies from the US or Europe. Though CAIN (CanArctic Inuit Networks) is an active participant in contemporary capitalism, as an Inuit-owned company, they reflect the long-held value in Nunavut of advancing Inuit economic sovereignty, a central element of the agreement established in the Nunavut Land Claims Agreement Act (NLCAA 1993).

To be clear, CAIN is not democratically or cooperatively owned. Owning and operating a cable system will generate profits, but the strong likelihood of necessary, expensive repairs means that cables will likely have slimmer profit margins compared to more easily managed and accessible forms of digital infrastructure, like data centers and satellites. Providing capacity is less profitable than other forms of infrastructure because there isn't a scarcity economic model in the same way. What seems to motivate CAIN is less profit, and more a genuine commitment to bring connectivity to Nunavut. SMART capacity, according to CAIN, is an extension of this connectivity, potentially bringing additional economic opportunity and situating Nunavut as vital in the region through the data they will be able to collect. SednaLink would connect Nunavut to the internet and to the environment in a new way. In interviews, Madeleine Redfern stated this clearly: "above all we need infrastructure that works" (Redfern, 2023). Connectivity infrastructure is particularly important to Redfern, who is committed to expanding the opportunities for Nunavut citizens, most of whom are Inuit (Redfern, 2023).

The efficacy of the cable, connectivity, and the SMART capacity of the SednaLink system and whether the system becomes a unique example of infrastructure is dependent on how the project is realized, along with the way CAIN and SednaLink interface with the community of Nunavut more broadly. Cable projects are not without controversy in the region. Many cable projects have been proposed in Northern Canada and most of those have failed or disappeared. Even projects that have been installed and launched service have faced significant challenges with maintenance because of the icy and dynamic geography across the region. Nunavut is no exception to this; the recent tabling of the KNF project is one example. These inconsistencies and failed projects are a continuation of the neglect of Nunavut and the rural areas of Northern Canada. Other forms of infrastructure are also urgently needed in the area which can lead to disagreements about where to direct investment. Without direct engagement with community members, which was beyond the scope of this chapter, it is unclear how SednaLink and CAIN are and will be received by the community. While SednaLink's fate and social, political, and environmental impact remain unclear, the potential of the project remains unique and points toward alternative possibilities for indigenous networks, collaborations between indigenous actors and science, and multi-purpose infrastructure that connects communities to both the global telecommunications network and environmental conditions.

SednaLink appears to serve as an example of what Marisa Duarte has called a sovereign network, or an infrastructural network operated by indigenous people that allows indigenous people to decide how that infrastructure interacts with the land and environment (Duarte, 2017). One of the ways this sovereignty is clear is through the choice of name. Through evoking the figure of Sedna, the SednaLink project gestures toward the ways the SMART cable project is entwined with the local and cultural knowledge of the Inuit people. Particularly, Sedna and the tradition surrounding her seem to speak to the ways in which the SednaLink project might be a site from which to imagine a kind of infrastructural remediation. The chance remains that the SednaLink system could simply reproduce extractive capitalism and existing geopolitical formations. Despite this, the historical legacy, cultural and geographic specificity of the figure and tradition of Sedna might also represent an alternative pathway forward for indigenous environmental and infrastructural futures.

Sovereignty and the Possibility of Infrastructural Remediation

Sedna is a central figure in the culture of the Inuit people.[8] Among the Inuit communities of Nunavut, she has been known as Nuliajuk and she is sometimes referred to simply as the Sea Woman, but different versions of the oral tradition and names of the figure appear across the North Atlantic in the Western hemisphere (Laugrand and Oosten, 2008). Sedna is the name that has become popularized, likely because of its syntactic and gendered legibility to Western readers and speakers. I will refer to her as Sedna because this is the name chosen for the SMART cable infrastructure that will connect Nunavut, a choice that potentially reflects the company's desire to be legible to the world beyond Nunavut. The oral tradition of Sedna was passed from generation to generation through the storytelling tradition of the Inuit people and varies across time and specific communities. As a central figure of the Inuit culture, Sedna's story informs the ways in which the Inuit communities understand the world, particularly the ocean. In a number of traditional iterations, Sedna's narrative is a kind of origin story for humanity and the creatures of the sea (Laugrand and Oosten, 2008, 2010; Canadian Museum of History, 2023). Though there are many versions, the stories share similar narrative points. Often, Sedna is a fierce, independent young woman who refuses to marry despite the wishes of her father. Her marriage is important, as a potential husband could help to secure food as an additional hunter. Sedna refuses all suitors until an unknown handsome hunter comes to her community. The handsome hunter eventually convinces Sedna to come with him to his home. Later, the narrative reveals the handsome hunter to be a trickster figure, either a dog disguised as a human or a bird spirit. In the versions of the tradition in which the handsome hunter is revealed to be a dog, Sedna gives birth to children that are part human, part dog. In the versions where this appears, legend says that children are said to be the origin of the different races of the world. In most versions, the handsome trickster takes Sedna to an island and her father comes to find her or otherwise ends up with Sedna near the ocean.

What happens next is the most significant element of the tradition and is consistent across all versions. Either because of a storm, or to save himself, or to punish her, Sedna's father attempts to throw her into the ocean. She holds onto either a cliff or a boat's edge, and he cuts off her fingers or her limbs. Each of her body parts becomes a different sea creature after they fall into the sea, including seals, sea lions, and whales, collectively representing the animals on which Inuit communities were most dependent. Eventually, she falls into the ocean and descends into the deep where she is transformed into the Sea Woman, protector of the sea and sea animals and a kind of representation of the ocean itself. As Sedna has no fingers, she is unable to comb her hair in the deep sea or otherwise take care of her body. An important ritual related to Sedna revolves around helping her to comb her hair, to relieve her of that pain and suffering in order to appease her and ensure a calm ocean (Angmaalik et al., 1999; Laugrand and Oosten, 2008). Crucially, the violence that Sedna faces in the story makes her vengeful and spiteful, or at least mercurial. She is certainly not a merciful figure; rather, she is a figure for whom the Inuit people are taught to fear and approach with reverence. Sedna requires appeasement. If Sedna is adequately cared for, appreciated, and appeased, she provides smooth waters for traveling and plenty of sea animals to fish and hunt. If not, it is believed that Sedna will produce storms and withhold sea animals to punish the Inuit people, disrupting the Inuit people's ability to travel or accumulate food and resources (Laugrand and Oosten, 2008).

As mentioned above, SMART cables cannot be divorced from the broader colonial histories of subsea cables as a key piece of colonial infrastructure. Infrastructure broadly has long been tied to colonialism and imperialism through the ways infrastructure is strategically deployed to arrange hegemonic forms of management, control, information, and economic exchange, particularly in relation to indigenous communities. Infrastructure has historically mediated relationships in colonial territories, including relationships between people, technological institutions, and between local communities and the environment, a strategy referred to as "infrastructural mediation" (Ruiz, 2021). Infrastructural mediation has been a key feature in the process of settler colonialism across the globe, with the design of developing infrastructure that undermines and devalues the practices, culture, and knowledges of local, indigenous communities. Subsea cables are no exception to this process; undersea cables have always been embedded and entangled in geopolitics and colonialism (Hurd, 1899; Kennedy, 1971; Starosielski, 2015). Likewise, as environmental sensing technology, SMART cables are enmeshed in the contemporary data-driven capitalism, practices of data extraction, and the emergent regime some have called data colonialism (Couldry and Mejias, 2019; Sadowski, 2019; Srnicek, 2017; Zuboff, 2015). Similar sensing projects, particularly associated with "smart cities," deployed across Canada, including in Nunavut, have been critiqued for the cultures of surveillance and data extraction (Valverde and Flynn, 2020).

Figure 11.2 An etching by artist Germaine Arnattauyuq depicting Sedna and her children, the sea creatures[9]
Source: Collection of the Winnipeg Art Gallery.

Despite the intertwined histories of infrastructure and colonialism, or maybe because of these histories, infrastructure is significant to reconsider the ways technological and environmental imaginations and relations are mediated and the subsequent worlds produced by infrastructural mediation. Infrastructural networks and systems are not inherently colonialist and extractive. Rather, their emergence and their specific deployment situate them within certain epistemic and political regimes. Indigenous studies scholars have recently argued that infrastructural remediation can manifest as a form of indigenous resistance or worldmaking, an argument that resonates with Duarte's argument for network sovereignty (Duarte, 2017). Winona LaDuke and Deborah Cowen effectively argue for reimagining infrastructure that is life-giving rather than the self-cannibalizing infrastructure of settler colonialism and capitalism (LaDuke and Cowen, 2020, p. 245). Jordan Kinder argues for the potentiality of local, indigenous solar projects in western Canada as decolonial, sovereign energy infrastructures, or "indigenous solarities" (Kinder, 2021). These arguments point to infrastructure's capacity to contribute to a remediation of the environment and economy that produce new social and epistemological arrangements that complicate and even potentially push beyond hegemonic, colonial logics and structures. While infrastructures are not inherently liberatory, they are uniquely situated to reconceptualize categories of understanding and producing alternative social formations.

The rhetoric of naming is often deployed as a cynical gesture to mask certain ideological commitments or to create the illusion of certain values. At the same time, naming can be a rhetorical gesture used to honor and commend, to memorialize. In the subsea cable industry, the names of cables are chosen according to two dominant logics, with some exceptions. A cable is named for the points it connects or the ocean a cable crosses. In the latter case, the cable is given a sequential number as part of the name for each updated version of the cable.[10] Notably, the SednaLink cable does neither. Instead, the name of the cable is suggestive of the particular, culturally specific role of the figure of Sedna in Inuit life. CAIN representative Madeleine Redfern stated that the choice was inspired by Sedna's role as a central figure of Inuit understanding and engagement with the ocean and in the Inuit models of resource exchange, including harvesting food and transportation (Redfern, 2023). Sedna is a figure tied to both environment and economy. The narrative of Sedna can be understood as detailing her emergence and ongoing role as an infrastructure for the Inuit community, if infrastructure is to mean that which connects us to each other, the environment, and the rest of the world as it has been broadly theorized (Berlant, 2016; Durham Peters, 2015; Parks, 2015). In one way, Sedna is a kind of allegory for the SednaLink through the kinds of connections they both facilitate. Sedna connects the Inuit people to critical resources and the oceanic environment, just as SednaLink does through its dual capacities. SednaLink will not inspire the kind of fear and reverence the tradition of Sedna demanded, though SednaLink will make the people of Nunavut aware of environmental harms brought about by events like climate change or geologic movement requiring a different, but similar kind of awareness and reverence. However, in thinking of Sedna as an infrastructure, it is possible to think beyond allegory to consider the ways Sedna serves as a social and epistemic model for the eventual role of SednaLink. Sedna provides access to transportation and hunting, to the environment and economy, and is capable of removing or restricting such access. Through the ongoing awareness and attention to the oceanic conditions and environment of the ocean, Sedna represents a kind of sensing practice. The Inuit communities of Nunavut have engaged in sensing practices independent of Sedna, but Sedna's presence and her connection to the deep sea required a particular awareness of oceanic conditions, one that has been interpreted and circulated across communities. Sedna as a figure presents a unique form of sensing – one that is deeply felt, affective, relational, embodied. She produces environmental effects and affects with consequences for the human and nonhuman beings with which she is always in relation. These effects and affects then produce certain forms of environmental and social action.

The choice to name the SednaLink after Sedna is suggestive of the kind of remediation offered by the cable system. The tradition of Sedna foregrounds the significance of awareness, responsiveness, and reciprocity with the oceanic environment in order to gain access to the resources of the ocean and to be protected from its potentially dangerous conditions. In drawing out this connection to the tradition of Sedna, environmental awareness and reciprocity in particular, SednaLink may represent the possibility of infrastructural remediation where subsea infrastructure becomes a point of connection between indigenous traditions and science. Rather than a colonial-deployment-as-management tool, one may argue that SednaLink is an infrastructure that has the potential to emphasize and reflect the culture of the Inuit people while advancing their economic and environmental sovereignty. The connection this chapter attempts to draw out is not purely metaphorical. SednaLink is in line with existing efforts in Nunavut to utilize traditional knowledges to address environmental issues, like climate change (Cameron, Mearns, and Tamalik McGrath 2015). Traditional practices of hunting and fishing are still utilized in Nunavut and as such traditional practices for knowing and observing the land, water, and ice are still important, but also increasingly under threat by shifting environmental dynamics (Nunavut Climate Change Centre, 2011). SednaLink's name is significant for the ways it signals a continuation of these traditions and forms of awareness and attention to the shifting dynamics of the ocean, particularly in light of climate change and its consequences. Traditional practices and contemporary science are both necessary to address these changes and the potential threat to Inuit life produced by climate change. As one Inuit leader wrote in an essay published by the UN, "The arctic ocean and the sea ice is our *nuna*," or home (Eegeesiak, 2017). SednaLink could represent an infrastructure that helps to protect the indigenous lands and waters of the Inuit while providing them effective access to the global network beyond Nunavut.

While a democratically-owned and operated internet infrastructure arguably would be a more ideal political formation, SednaLink could serve as a model for other existing and emergent infrastructures to be reimagined to emphasize the forms of traditional ecological and site-specific knowledge present in oral traditions, along with the scientific knowledge. Indigenous scholars from a range of disciplines have called for the inclusion of traditional ecological knowledge into our approaches to the environment, including scientific practices, as well as infrastructure that emphasizes or is informed by traditional ecological knowledge and indigenous ways of knowing. In particular, in light of the climate crisis and the emerging, intersecting ecological crises of our contemporary world, scholars have called for a broader consideration of traditional ecological knowledge and indigenous ways of knowing to adequately address ecological crises and to produce ecologically sustainable and socially just futures. Indigenous philosopher Kyle Whyte has written extensively about producing climate futures rooted in indigenous knowledge and sovereignty (2017, 2013). Indigenous philosopher and scientist

Robin Wall Kimmerer has called for the introduction of more traditional ecological knowledge into environmental science practices (2013). Scholars across media studies have called for the necessity of indigenous sovereignty with respect to infrastructure and infrastructural networks, most notably Marisa Duarte.[11] Network and infrastructural sovereignty are necessary precisely for the ways in which they produce tensions across indigenous modes of knowing and relating to the environment, technology, and culture against settler colonial networks, infrastructures, and logics and broader geopolitical formations. As scholars like LaDuke, Cowen, and Kinder have argued, the production of sovereign indigenous infrastructures establishes the ground for remediating relations across environments in ways that de-emphasize colonialist logics and allow for different kinds of environmental futures (Kinder, 2021; LaDuke and Cowen, 2020).

SednaLink is an infrastructure that inherits the histories of colonialism and empire and the ways these ideologies manifest in the technological reality of subsea infrastructure. But through its deployment, the system has the potential to exceed the extractivist and colonial frameworks with which the technology is associated. Though SMART cables are a technology that has been adopted or will be deployed in different contexts around the world, SednaLink may demonstrate that an infrastructure and technology can be reimagined according to the site-specific contexts, the cultural and ecological specificity, of its operation. Furthermore, it could point to the much broader potential of infrastructure to produce different models of understanding that challenge, undermine, or resist colonial expectations. This is particularly significant for considering the ways in which indigenous communities take up infrastructures that have colonial associations and histories in order to gain access to the global economic landscape, especially as they do so in ways that are expressions of their local culture, traditions, ecological practices, and sovereignty. Reimagining and reconfiguring infrastructures and the worlds they help to construct is a critical task in light of the climate crisis and the geopolitical landscape that climate crisis will continue to usher in through its material effects. This certainly seems true in the High North Atlantic, an area of the world where infrastructure is critical to address the unique environmental challenges faced by proximity to the Arctic—environmental challenges that will persist, shift, and intensify as climate crisis continues to unfold.

Notes

1 Madeleine Redfern is the COO of CanArctic Inuit Networks and a leading activist and policy advocate for expanding connectivity in Nunavut. I interviewed her for this project as she is a lead project manager on the development of the SednaLink subsea system.

2 See, https://www.oceannetworks.ca/; See DONET: Mulia, Iyan E., and Kenji Satake. "Developments of Tsunami Observing Systems in Japan." *Frontiers in Earth Science*, 8(145), 2020.

3 See, https://www.oceannetworks.ca/; See DONET: Mulia, Iyan E., and Kenji Satake. "Developments of Tsunami Observing Systems in Japan." *Frontiers in Earth Science*, 8(145), 2020.
4 More on UN Joint Task Force on SMART cables here: https://www.itu.int/en/ITU-T/climatechange/task-force-sc/Pages/default.aspx.
5 CanArctic Inuit Networks is registered in the Inuit Firm Registry Database, a publicly available database that tracks Inuit-owned businesses to highlight and advance Inuit economic sovereignty, "CanArctic Inuit Networks Inc.," https://inuitfirm.tunngavik.com/inuit-firms/canarctic-inuit-networks-inc/, 2023. The IFRD declares CAIN as a level 1 company, meaning Inuit ownership stake is between 50 and 75 percent.
6 Information regarding these acts can be found here: Nunavut Land Claims Agreement, https://lois-laws.justice.gc.ca/eng/acts/N-28.7/FullText.html, and Nunavut Act, https://laws-lois.justice.gc.ca/eng/acts/N-28.6/index.html
7 See Nunavut Tunngavik Inc.'s report "Nunavut's Infrastructure Gap." NTI is an organization that represents the territory's 33,000 Inuit and their rights under the Nunavut Agreement.
8 One of the most popular and well-known Inuit oral traditions, Sedna has become widely circulated as an Inuit narrative in contemporary times, including as the inspiration for the name of a planet discovered in 2003, the subject of an animated film for children, and as a featured character in an issue of the comic *Aquaman*. Sedna's importance to the Inuit cultures of the North Atlantic and Arctic is most easily traced through her ubiquity in the region, her regular appearance in Inuit art, and her role in the cosmology of the Inuit people.
9 Germaine Arnattauyuq, Sedna, Ruler of All Sea Animals, 1994. Collection of the Winnipeg Art Gallery, Gift of Moira Swinton and Bernard Léveillé.
10 For more on the names of cables, consult submarinecablemap.com, a map produced by Telegeography and Aquacomms.
11 Marisa Duarte, *Network sovereignty: Building the internet across Indian country*, University of Washington Press, 2017, p. 25.

References

Arctic360 (2022). Episode 14: The future of internet in the North with Madeleine Redfern. *Arctic360*, https://arctic360.org/episode-14-the-future-of-internet-in-the-north.

Angmaalik, P., Nakasuk, S., Ootoova, E., and Paniaq, H. (1999). *Vol. 1 of Interviewing Inuit Elders*. Frédéric Laugrand and Jarich Oosten (Eds.). Arctic College.

Bell, J. (2021). New undersea fibre-optic pitch for Nunavut competes against territorial government's plans. *Nunatsiaq News*. https://nunatsiaq.com/stories/article/new-undersea-fibre-optic-pitch-for-nunavut-competes-against-territorial-governments-plans/.

Berlant, L. (2016). The commons: Infrastructures for troubling times. *Environment and Planning D: Society and Space*, 34(3), 393–419.

Butler, R. (2012). Using submarine cables for climate monitoring and disaster warning strategy and roadmap: A report. ITU. https://www.itu.int/dms_pub/itu-t/oth/0b/25/T0B250000013302PDFE.pdf.

Cameron, E., Mearns, R., and McGrath, J.T. (2015). Translating climate change: Adaptation, resilience, and climate politics in Nunavut, Canada. *Annals of the Association of American Geographers*, 105(2), 274–283.

CanArctic Inuit Networks (2021). CanArctic Inuit Networks' SednaLink Fibre to eliminate Nunavut and Nunatsiavut Connectivity Crisis by November 2022, *SubTel Forum*. https://subtelforum.com/fibre-optic-network-between-iqaluit-nu-a

nd-clarenville-nl-which-will-dramatically-improve-connectivity-in-to-inuit-nunangat-by-november-2022/.

Canadian Museum of History (2023). Featured works of art: Human dog spirits in slipper. *Canadian Museum of History.* www.historymuseum.ca/cmc/exhibitions/aborig/iqqaipaa/ex57e.html.

Couldry, N., and Mejias, U.A. (2019). Data colonialism: Rethinking big data's relation to the contemporary subject. *Television & New Media*, 20(4), 336–349.

CRTC (2020). Communications monitoring report. CRTC.

Duarte, M. (2017). *Network sovereignty: Building the internet across Indian country.* University of Washington Press.

Eegeesiak, O. (2017). The Arctic Ocean and the sea ice is our nuna. *UN Chronicle.* https://www.un.org/en/chronicle/article/arctic-ocean-and-sea-ice-our-nuna.

Flynn, A. (2021). The poverty of broadband infrastructure in Nunavut can and must be ended. *Centre for Free Expression Blog.* https://cfe.torontomu.ca/blog/2021/10/poverty-broadband-infrastructure-nunavut-can-and-must-be-ended.

Gabrys, J. (2015). *Program earth: Environmental sensing technology and the making of a computational planet.* University of Minnesota Press.

Haight, M., Quan-Haase, A., and Corbett, B.A. (2016). Revisiting the digital divide in Canada: The impact of demographic factors on access to the internet, level of online activity, and social networking site usage. In J. Earl and K. Kimport (Eds.), *Current research on information technologies and society* (pp. 113–129). Routledge Press.

Halpern, O. and Mitchell, R (2023). *The smartness mandate.* MIT Press.

Howe, B.M., Arbic, B.K., Aucan, J., Barnes, C.R., Bayliff, N., Becker, N.*et al.* (2019). SMART cables for observing the global ocean: Science and implementation. *Frontiers in Marine Science*, 6(424). doi:10.3389/fmars.2019.00424.

Howe, B.M., Angove, M., Aucan, J., Barnes, C.R., Barros, J.S. et al. (2022). SMART subsea cables for observing the earth and ocean, mitigating environmental hazards, and supporting the blue economy. *Frontiers in Earth Science*, 9, 775544.

Hurd, A.S. (1899). An all British cable system, *The Nineteenth Century: A Monthly Review, Mar. 1877–Dec. 1900*, 45(264), 226–234.

Kennedy, P.M. (1971). Imperial cable communications and strategy, 1870–1914. *English Historical Review*, 86(341), 728–752.

Kimmerer, R.W. (2013). *Braiding sweetgrass: Indigenous wisdom, scientific knowledge and the teachings of plants.* Milkweed Press.

Kinder, J.B. (2021). Solar infrastructure as media of resistance, or, indigenous solarities against settler colonialism, *South Atlantic Quarterly*, 20(1), 63–76.

LaDuke, W., and Cowen, D. (2020). Beyond Wiindigo infrastructure. *South Atlantic Quarterly*, 119(2), 243–268.

Laugrand, F., and Oosten, J. (2010). *Inuit Shamanism and Christianity: Transitions and transformations in the twentieth century.* McGill–Queen's University Press.

Laugrand, F., and Oosten, J. (2008) *The SeaWoman: Sedna in Inuit Shamanism and art in the eastern Arctic.* University of Alaska Press.

Lochead, D. (2023). In Nunavut, Starlink stands up to cold but service, speeds vary, users say. *Arctic Business Journal.* https://www.arctictoday.com/in-nunavut-starlink-stands-up-to-cold-but-service-speeds-vary-users-say/.

Lochead, D. (2022a). Nunavut fibre optic plans face new delays. *Nunatsiaq News.* https://nunatsiaq.com/stories/article/nunavut-fibre-optic-plans-face-new-delays/.

Lochead, D. (2022b). Starlink delivers fast internet, some outages, in Nunavut. *Nunatsiaq News*. https://nunatsiaq.com/stories/article/nunavummiut-report-blazing-speeds-temporary-outages-with-starlink.

Durham Peters, J. (2015). Infrastructuralism: Media as traffic between nature and culture. In M. Näser-Lather and C. Neubert (Eds.), *Traffic: Media as infrastructures and cultural practices* (pp. 31–49). Brill Press.

Nunatsiaq News (2021). Comments section of 'New Undersea Fibre-optic Pitch for Nunavut competes Against Territorial Government's Plans'. *Nunatsiaq News*. https://nunatsiaq.com/stories/article/new-undersea-fibre-optic-pitch-for-nunavut-competes-against-territorial-governments-plans/.

Nunatsiaq News (2022). Comments section of 'Nunavut fibre optic plans face new delays'. *Nunatsiaq News*. https://nunatsiaq.com/stories/article/nunavut-fibre-optic-plans-face-new-delays/.

Nunavut Land Claims Agreement Act (1993). https://laws-lois.justice.gc.ca/eng/acts/N-28.7/page-1.html.

Nunavut Climate Change Centre (2011). Upagiaqtavut Setting the Course: Climate Change Impacts and Adaptation in Nunavut. *Nunavut Climate Change Centre*. https://climatechangenunavut.ca/sites/default/files/3154-315_climate_english_reduced_size_1_0.pdf.

Nunavut Tunngavik Inc. (2020). *Nunavut's Infrastructure Gap. Iqaluit: Nunavut Tunngavik Inc*. https://www.tunngavik.com/files/2020/10/2020.10.20-Nunavuts_Infrastructure_Gap_Report_vf.pdf

Parks, L. (2015). Stuff you can kick: Toward a theory of media infrastructures. In D.T. Goldberg and P. Svensson (Eds.), *Between Humanities and the Digital* (pp. 355–373). MIT Press.

Peters, J.D. (2015). *The marvelous clouds: Towards a philosophy of elemental media*. University of Chicago Press.

Redfern, M. (2023). Interview with Madeleine Redfern. Conducted by Nicholas Silcox.

Ruiz, R. (2021). *Slow disturbance: Infrastructural mediation on the settler colonial resource frontier*. Duke University Press.

Sadowski, J. (2019). When data is capital: Datafication, accumulation, and extraction, *Big Data & Society*, 6(1).

SednaLink Marine Systems (2025). *SednaLink Marine Systems. Crunchbase*. https://www.crunchbase.com/organization/sednalink-marine-systems.

Srnicek, N. (2017). *Platform capitalism*. John Wiley & Sons.

Starosielski, N. (2020). Depth mediators: Undersea cables, Network infrastructure, and the deep ocean. In K. Redrobe and J. Scheible (Eds.), *Deep mediations: Thinking space in cinema and digital cultures* (pp. 262–285). University of Minnesota Press.

Starosielski, N. (2021). *Media hot and cold*. Duke University Press.

Starosielski, N. (2015). *The undersea network*. Duke University Press.

Submarine Cable Network (2021). CanArctic Inuit Networks and Norway's bulk fiber networks collaborate on subsea fiber-optic network into Nunatsiavut and Nunavut. *Submarine Cable Networks*. https://www.submarinenetworks.com/en/systems/north-america/sednalink/canarctic-inuit-networks-and-norway-s-bulk-fiber-networks-collaborate-on-subsea-fiber-optic-network-into-nunatsiavut-and-nunavut.

Taqqut Productions (2002). The legend of Nuliajuk. *CBC North Radio One: Inuit Legends: Volume 1*. CBC.

Telegeography. (2021). The state of the network. https://www2.telegeography.com/hubfs/assets/Ebooks/state-of-the-network-2021.pdf.

Tranter, E. (2023). Nunavut cancels tender for project that would've brought high speed internet to Iqaluit, Kimmirut. *CBC News*. https://www.cbc.ca/news/canada/north/nunavut-cancels-tender-for-fibre-optic-line-internet-1.6900346.

Valverde, M. and Flynn, A. (2020). *Smart cities in Canada: Digital dreams, corporate designs*. Lorimer Press.

Whyte, K. (2017). Indigenous climate change studies: Indigenizing futures, decolonizing the Anthropocene. *English Language Notes*, 55(1), 153–154.

Whyte, K. (2013). On the role of traditional ecological knowledge as a collaborative concept: A philosophical study. *Ecological Processes*, 2(1).

Wright, T. (2023). CanArctic signs MOU with Alaskan company to bring fibre to Iqaluit. *Nunavut News*. https://www.nunavutnews.com/news/canarctic-signs-mou-with-alaskan-company-to-bring-fibre-to-iqaluit/.

Zuboff, S. (2015). Big other: Surveillance capitalism and the prospects of an information civilization. *Journal of Information Technology*, 30(1), 75–89.

12 Ocean Futures in the Making

Camilla Brattland and Synne Tollerud Bull

Introduction

The video and sound installation, *Acoustic ocean* (2018) by artist, author, and video essayist, Ursula Biemann, is set up as an immersive and evocative experience for the visitor.[1] The installation places the viewer within a large video projection, where the dark hum of underwater reverberations merges with the rhythmic crash of distant waves, establishing a powerful sonic and visual maritime atmosphere. The video in the installation presents an ocean framed by jagged, barren mountains and a stark shoreline, where the entanglement of hydrophones and cables, laid out like tentacles across the rocks, capture the murmurs of unseen marine life. The figure—referred to by Biemann as an "aquanaut"—dressed in a bright orange diving suit with a reindeer-skin neckpiece—moves methodically along the water's edge, calibrating her instruments. The installation's audio track presents an ocean teeming with signals, where clicks and whirs merge with the distant calls of marine mammals, weaving a soundscape that conveys both ecological vitality and vulnerability. Originally commissioned by the Atlantic Project: *After the Future* in Plymouth, UK, the work explores oceanic ecologies of Indigenous knowledge embedded as a potential part of the Lofoten archipelago, a part of the High North Atlantic known for its rich fisheries and coastal cultural heritage.[2] Central to the project is Biemann's collaboration with Sámi performer and climate activist, Sofia Jannok, whose engagement shapes the work's approach to merging Indigenous ecological knowledge with modern sensing technologies. In *Acoustic ocean*, Jannok plays the role of aquanaut or Indigenous marine scientist, a hybrid between Indigenous ecologist and marine scientist, clad in reindeer fur, with a scientific approach to sensing the ocean. What kind of alternative values and futures does she represent, and what ecological future alternative does the work of art offer to its viewers?

Located above the Arctic Circle in Northern Norway, the Lofoten, Senja, and Vesterålen archipelagos extend prominently into the Norwegian Sea, northwest of Norway between the North Sea and the Greenland Sea, adjoining the Barents Sea to the northeast. The cluster of islands act as a natural barrier that disrupts marine currents, inducing the upwelling of

DOI: 10.4324/9781003478805-12

nutrient-rich deep waters that sustain the region's highly productive ecosystems. They serve as critical spawning grounds for the world's largest population of North-East Arctic cod (Gadus morhua), known as *skrei*, and host the North Atlantic's largest nesting sites for seabirds and whales, along with the world's most extensive cold-water coral reef. Closer to shore lie significant oil reserves, while further out—some 250 nautical miles offshore—areas are being mapped as potential frontiers for deep-sea mining. Consequently, the region is subject to ongoing tensions between conservation and resource exploitation. In recent years, Norway has taken steps to open parts of its extended continental shelf for deep-sea mineral extraction, claiming this is necessary to secure critical resources for renewable energy technologies (Norwegian Ministry of Energy, 2024). However, this decision remains contentious, with environmental concerns delaying the first mining licenses (Reuters, 2024). As Kristoffersen et al. (2021) suggest, such debates reflect broader tensions between temporalities, with oil extraction and traditional community life embodying two vastly different futures for the region. In the current political climate, renewed interest in oil exploration off the coast of Lofoten, Vesterålen, and Senja—particularly amongst younger politicians—suggests that an oil-driven future for the region could be back on the agenda sooner than we think (Staberg and Budalen, 2024). In *Acoustic ocean*, the aquanaut moves through a seascape that appears both familiar and estranged, as if searching for traces of life in an ocean altered by extractive industries. The work gestures toward a speculative future where the impacts of human activity are deeply inscribed in the marine environment, raising questions about what forms of life and knowledge persist and what ecological alternatives remain possible.

To those familiar with the history of the Lofoten area and Northern Norway, the Indigenous aquanaut appears an unusual, somewhat puzzling figure. The typical image is that of the male Norwegian fisherman clad in orange oilskins, as seen in photographs by Anders B. Wilse from the early 20th century, while Sámi presence in the region has been comparatively silenced in historical and scientific narratives. The longstanding customs of Northern Scandinavia's Indigenous communities—shaped through close interaction with fragile ecosystems—represent a temporality that, as Kristoffersen et al. (2021) suggest, remains underexplored in current debates on the value of marine ecosystems, especially when compared with extractive temporalities shaping current development agendas. As Alf Ragnar Nielsen contends in his book, *Samisk Fortid i Lofoten* (Nielsen, 2024), the fishing tradition in the Lofoten archipelago has been integral to the region's cultural and economic identity for over a millennium, its origins tracing back to the Viking Age. This maritime heritage remains a cornerstone of Lofoten's contemporary society, symbolizing the enduring relationship between the inhabitants and the vast marine environment that has shaped their livelihoods and traditions over centuries. A lesser-known part of this region's story, however, is the coastal Sámi population and their ancestral knowledge of oceanic relations and cohabitation—knowledge that, while often obscured in dominant narratives, is reactivated in Biemann's work as a form of speculative continuity. In the role of Indigenous scientist, Jannok channels this reimagined memory through carefully orchestrated technological mediation, revealing the hidden sounds and images of a vulnerable oceanic environment.

This chapter explores how Biemann's *Acoustic ocean* engages with these temporal tensions by positioning Jannok as a mediator between Indigenous knowledge and modern scientific methods, between present ecological choices and speculative futures. It asks: how can artistic research—understood as a mode of knowledge production (Borgdorff, 2012)—re-center marginalized perspectives and suggest alternative ecological futures for the High North Atlantic through multisensory engagements? Drawing on Jennifer Gabrys' (2019) concept of sensing practices and Jussi Parikka's (2015) notion of *medianatures*, the chapter explores how *Acoustic ocean* engages with extractive paradigms by opening toward alternative futures rooted in multispecies entanglement and Indigenous resilience. Gabrys and Parikka offer frameworks for thinking about how sensing technologies shape environmental perception—foregrounding the entanglement of human and nonhuman agencies, and how media infrastructures participate in ecological relations. Positioned within a wider field of environmental media theory and speculative Indigenous futurisms, this framework allows for an exploration of how *Acoustic ocean* navigates the intersection of Sámi ecological knowledge and modern scientific methods—not only through embodied sensing and technological mediation, but also through its speculative engagement with the figure of the Indigenous scientist. By weaving these perspectives together, this chapter contributes to the environmental humanities by demonstrating how artistic research can re-center marginalized knowledge systems and suggest inclusive, sustainable approaches to human-ocean relations amid the ongoing ecological crises.

Envisioning Ocean Futures in the Making

The region of Lofoten, Vesterålen, and Senja is estimated to contain significant untapped oil and gas reserves.[3] According to estimates from the Norwegian Petroleum Directorate (NPD), the total recoverable resources in the area are believed to be between 1.3 and 3 billion barrels of oil equivalent (boe). These estimates cover both oil and natural gas deposits. Since 2009, the People's Action for an Oil-free Lofoten, Vesterålen, and Senja (LoVeSe), consisting of approximately 5,000 members, has united environmental activists, fishermen, and local communities—both young and old—in efforts to protect the region from oil exploration. With support from the Institute of Marine Research, the group campaigned to keep oil extraction out of the region. In the fall of 2023, the action group officially concluded its activities, declaring its mission accomplished, with the Norwegian Parliament (Stortinget) having issued no new oil licenses in the area. However, local and environmentalist actors in the Lofoten region remain committed to its green and blue profile as a rich natural and cultural heritage area working towards sustainable tourism under the label of "the green islands." Several processes aimed at achieving World Heritage status are under way, while the region is also a candidate for the status of Biosphere Reserve under the UNESCO Man and the Biosphere program. Its profile has thus far centered on an image of the region as an exclusively Norwegian area with little or no reference to its Sámi history or as an important region for Sámi culture.

The discourse on the development of the Lofoten and Vesterålen area has also remained almost clinically free of reference to the impacts of oil extraction on the "material basis" for Sámi and coastal Sámi culture and livelihoods. According to the Marine Resource Act of 2009, the living marine resources of Norway belong to the "people in common," which, in effect, also means that they are an important material basis for the culture and livelihoods of the peoples subsisting from these resources, including the Sámi. The lack of connection between marine resources and cold-water reefs, for this matter, is a striking contrast to the grassroots environmental movement against wind farming, where the negative impacts on reindeer and reindeer husbandry as a traditional Sámi way of life figure as one of the main factors of the movement. This was particularly illustrated by the Fosen case, where environmental and Sámi interests joined in public demonstration against the development of wind farms on reindeer pastures, culminating in a Supreme Court ruling in the fall of 2023, stating that the industrial development was in violation of human rights (Fjellheim, 2023). In the context of this chapter, this is particularly interesting, as the environmental movement against wind farming has many parallels to the People's Action for an Oil-Free LoVeSe, but without the alliance between Sámi youth and environmental activists that so characterized the Fosen demonstrations of 2022 and 2023. The main difference between the two movements is the absence of a clear case of oil development negatively impacting Sámi culture and traditional ways of life. Even though there are no clear-cut cases of Sámi livelihoods and families being forced to leave their traditional way of life or having their human rights infringed upon in the face of Atlantic Ocean development, the lack of focus and attention to the issue says something about how Sámi presences other than those conforming to preconceptions of traditional Sámi ways of life are being discursively silenced in the field of ocean development. The lack of allowance for the Sámi perspective in this field illustrates a general lack of knowledge of the presence of Sámi culture in the LoVeSe region, and our preconceptions of what Sámi culture is and how it is visible.

Historically, Sámi from the Lofoten region were known to be skilled boat builders and sea farers prior to the 1600s and participated along with other seafarers in the Lofoten fisheries every season (Nielsen, 2024). The technologies employed by fishermen were based on human senses and navigational skills, being culturally specific to the diverse groups of fishermen drawn to the region from a wealth of Norwegian coastal communities every year between January and April to participate in the Lofoten fisheries. The Sámi who participated in the fisheries were both from the region and from further away, mostly consisting of men equipped with wool, food, and other essentials provided by women in their home communities. This pattern persisted until the 1950s, when modernization and the motorization of the fishing fleet transformed it from relying on home-made materials and provisions to modern clothing and products bought from stores. Sámi women also participated in the fisheries, among them the renowned "Buks-Berret" or Beret Paulsdatter (1794–1868) who owned her own fishing boat and who, in 2022, had a new university research vessel named after her (Asdal and Huse, 2023). Beret earned her nickname for wearing trousers (bukser) instead of skirts.

Local and Indigenous knowledge of the ocean predates the first scientific explorations and publications on this part of the ocean, such as G. O. Sars' investigations of cod spawning in Lofoten, which were commissioned by the Norwegian Parliament as an attempt to explain the variability of the cod catch over time. In 1864, Sars validated fishermen's claims that cod roe floated freely in the waters, as opposed to other species, where eggs were typically found on the sea bottom (Asdal and Huse, 2023). Fisheries following the seasonal cycles of the cod had been based on Indigenous and local knowledge systems for hundreds of years prior to this "discovery," illustrating that an intimate knowledge of the interaction between fish species, ocean currents, and the sea bottom was required to navigate and find congregations of cod in the Atlantic waters. The possibility that a Sámi fisherwoman could have a scientific vessel named after her shines a light on the common understanding of Sámi, Sámi women, and their capabilities as seafarers and as scientists. Another such figure was imagined in the artwork, *Acoustic ocean* (2018)—as the emerging figure of the female Indigenous scientist.

In the Fosen case, the main figures making it to the media were young Sámi women and men, amongst whom the most prominent was Sámi artist, Ella Marie Hætta Isaksen.

The group were supporting the south Sámi reindeer herding communities in the Fosen area, where wind power plants had been built without the free, prior, or informed consent of the Sámi, violating the human rights of the reindeer herders. It is perhaps surprising that the main defender of the rights of the south Sámi was not a southern Sámi reindeer herder, but a young female Sámi artist from a different Sámi community. The LoVeSe Peoples' Action Group attracted a diversity of societal groups from across Norway, including, but not primarily driven by Sámi, who were all in support of an environment and area that few had a direct relationship with. The cause itself and the values it represents, however, hold global relevance as part of the ongoing global environmental crisis, and a case of preservation of valuable biodiversity in the face of energy development. This goes to show that alliances between Indigenous and environmental movements go beyond the specific locality where the conflict is taking place, creating a wider community and action beyond defending the environment against the adverse impacts of energy

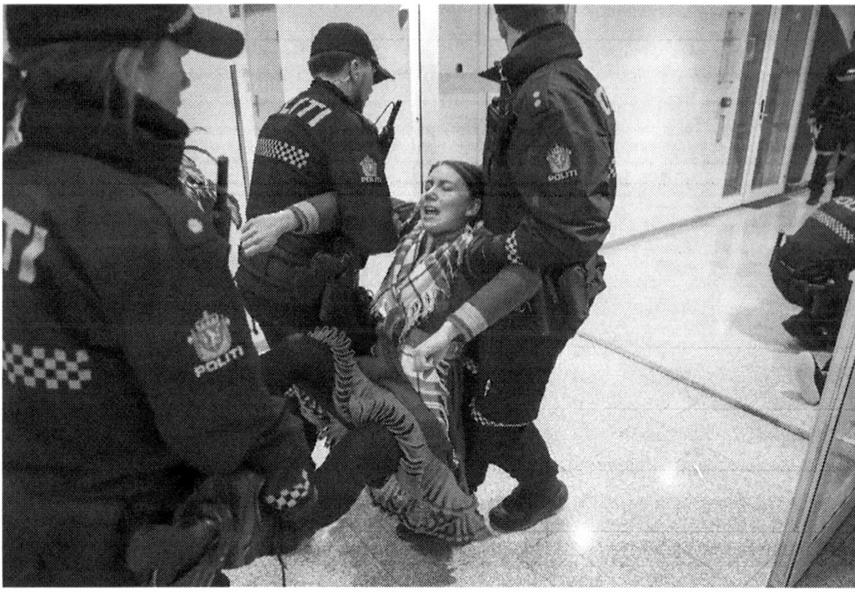

Figure 12.1 Ella Marie Hætta Isaksen during the Fosen demonstrations in Oslo, 2023. As one of the central figures in the protests, she drew national attention to the human rights violations facing Sámi reindeer herders
Source: Photo: Rasmus Berg / Nature and Youth Norway.

development, such as wind power plants in reindeer herding lands and oil extraction in important fish spawning or harvesting areas. Another important aspect lies in the figures and concepts emerging from and representing the connection between the environment and humans, such as the climate change activist, Greta Thunberg, and environmental and Indigenous rights defenders, Ella Marie Hætta Isaksen and Sofia Jannok.

One of the central aspects of ocean governance is that it is represented by the main active user groups of the ocean, namely fishermen, oil industry workers, and marine scientists, in addition to activists defending the environment. The Peoples' Action for an oil-free Lofoten, Vesterålen, and Senja was represented by a diverse group of northern activists with a majority of young Norwegians, the leading organization being Naturvernforbundet—Nature Defenders Norway (see the website folkeaksjonen.no and the book on the movement "Leve havet" [Long live the ocean] by Delsett, 2023). Among these representatives, however, there was a striking lack of Indigenous figures representing the marine environment and the Atlantic Ocean. Based on this disconnect, what can we imagine as explorers of the imagery of the High North Atlantic? To understand the disconnection between the dominating image of the marine environment and its role as a basis for Indigenous culture, we explore how Biemann's *Acoustic ocean* bridges disparate cultural images and epistemologies, offering pathways for engagement with the High North Atlantic.

Acoustic Ocean

The opening sequence of the video in Ursula Biemann's installation, *Acoustic ocean* immerses the viewer in a 3D-rendered deep blue oceanic environment, accompanied by a continuous, tension-building hum interspersed with distant acoustic signals. Intermittent textual overlays, appearing in place of a voice-over, deliver a silent narration that unfolds gradually throughout the scene, heightening its cinematic tension. The narration introduces the SOFAR (Sound Fixing and Ranging) channel, a deep-sea acoustic waveguide discovered in the mid-20th century.[4] This phenomenon occurs at a depth where temperature and pressure conditions create a low-sound-velocity layer, effectively trapping and guiding sound waves across vast distances with minimal energy loss. Discovered in the mid-20th century, the SOFAR channel was first used in military and oceanographic contexts, enabling long-range underwater communication and also serving as a key navigational aid for marine mammals. Both submarines and marine mammals rely on this natural waveguide for communication, but increasing anthropogenic noise pollution threatens its fragile acoustic balance, disrupting vital ecological interactions (Hildebrand, 2009). By foregrounding this natural underwater sound channel and its vulnerability to technological intrusion and ecological degradation, the work gestures toward how ocean sensing relies on more-than-human infrastructures of communication, shared by submarines and marine mammals alike.

Jennifer Gabrys' concept of *sensing practices* provides a key framework for understanding how environmental knowledge is shaped through situated and relational engagements between humans, technologies, and ecosystems (Gabrys, 2016, 2019). Rather than framing sensing as the passive registration of data, Gabrys foregrounds the ways in which infrastructures of measurement—ranging from satellites to DIY monitors—actively shape what becomes perceptible and politically actionable. Sensing, in this view, is a co-constitutive process entangled with ecologies of care, attention, and contestation. Complementing this perspective, Jussi Parikka's notion of *medianatures* highlights how media technologies are themselves ecologically embedded and materially contingent—dependent on extractive infrastructures while also conditioning how environments are sensed and understood (Parikka, 2015). Together, these concepts challenge anthropocentric and technocentric models of mediation by attending to the recursive entanglements between perception, matter, and environment. When the SOFAR channel is enrolled into hydro-acoustic monitoring systems, it becomes part of what Gabrys (2019) describes as an "infrastructure of perception"—where environmental sensing is not only about detecting phenomena, but also about shaping the very conditions under which relations between species, technologies, and environments are produced and perceived. This conceptual framing sets the tone for the film's multi-sensory mediation—both in form and in content—as it transitions to a rocky shore at dusk, where the jagged black peaks of the Lofoten archipelago are outlined against the diffuse, low-angle glow of the setting sun.[5]

As Biemann's video transitions to the next scene, the title *Acoustic ocean* dissolves into rippling waves, highlighting the shared wave properties across different forms and mediums. The camera remains fixed on the ocean waves, breaking white against the rocky shore with a semi-regular rhythm. A large, metal equipment box is centrally positioned within the frame, interrupting the continuity of the otherwise unspoiled seascape. Suddenly, a female character (Jannok) fades into view, as though she had always been present, choosing this moment to reveal herself. Referred to in *Acoustic ocean*'s narration as the aquanaut, she appears in an orange deep-sea diving suit with a reindeer-skin neckpiece—a hybrid figure that evokes a confluence of ancestral knowledge and speculative futures. In dialogue with the preceding discussion of the erasure of Sea Sámi presence from cultural memory, and, by extension, their absence from the popular environmental movement in the Lofoten, Vesterålen, and Senja regions, this character, a kind of "Indigenous scientist" figure, raises questions not only about epistemological relations, but also about memory, absence, and the reimagining of erased coastal life-worlds. For viewers embedded in local maritime traditions, the aquanaut's orange diving suit may further recall the iconic oilskins of Lofoten fishermen—a visual resonance that complicates the work's layered engagements with memory and erasure. In the video, its bright color and headlights stand out against the otherwise blue-tinted atmosphere. Jannok's gradual emergence and composite appearance recall what Michelle Raheja (2010) calls a "visual sovereignty," asserting an Indigenous presence within the speculative and technological futures from which it has historically been excluded. Jannok's appearance thus not only asserts Indigenous presence, but also exemplifies a wider aesthetic strategy in which Indigenous figures intervene in dominant scientific and technological imaginaries.

This assertion of Indigenous presence aligns with a broader tendency in contemporary art to reconfigure scientific authority through Indigenous and feminist epistemologies—a tendency Biemann's staging of the aquanaut in *Acoustic ocean* clearly embodies. Scholars such as Grace Dillon (2012), Michelle Raheja (2010), and Suzanne Kite and Jason Lewis (2021) have shown how the figure of the Indigenous scientist or technologist becomes a vehicle for imagining futures grounded in ancestral knowledge systems—not as vestiges of the past, but as speculative orientations toward planetary entanglements. In this context, technology functions as an extension of land- and ocean-based epistemologies. *Acoustic ocean* resonates with this artistic and scholarly momentum, particularly in its deployment of sensing technologies as both narrative device and epistemological proposition. What kinds of epistemological and ecological frameworks are mobilized here, and how might they reshape dominant understandings of marine perception, governance, and cohabitation? The analysis that follows explores how such refigurations unfold, and what forms of multispecies imaginaries and political sensibility they bring into view.

As the scenes in *Acoustic ocean* unfolds, the aquanaut opens the box and carefully unpacks a collection of remote acoustic sensing equipment, arranging the instruments along the rocky terrain. Cables are unfurled, hydrophones positioned, and dials adjusted with methodical precision. Once everything is in place, she begins tuning in with her headphones, and the film's soundtrack seamlessly transitions from the familiar rhythm of ocean waves to an array of unfamiliar sonic textures. Squeaks, whirs, and clicks reveal the intricate acoustic milieu of a hidden subaquatic world. Another shot from the shore toward the ocean introduces a second sequence of silent on-screen text, unfolding without voice-over. These describe the aquanaut's actions in poetic terms, foreshadowing the gestures she will later perform on screen. When she submerges hydrophones into the ocean, it is described as an act of intercepting "the vocal signals marine beings are sending through." The acoustic instruments are referred to as external organs that enable her "to deeply immerse herself in the aquatic habitat," allowing her to adjust "the frequencies to capture the dense sonic signature of entire ecosystems." This choice of wording reinforces the aquanaut's role as more than a technician; she becomes an intermediary within a distributed sensing system, where agency is shared between human, technological, and marine actors. The notion of "external organs" resonates with Indigenous epistemologies as articulated by Kite and Lewis (2021), who emphasize how technological extensions of the body can be grounded in relational practices with land, sea, and more-than-human life. By shifting from an attentive to an immersive mode, the aquanaut enacts a mode of knowing in which sound is not only received but co-shaped—disclosing, as the narration states, "a sea full of intentions." Here we can see an example of Gabrys' (2019) concept of a relational and situated sensing practice—one in which technologies are highlighted as active participants in shaping how environments become intelligible.

The poetic language of the narration reinforces the film's portrayal of the aquanaut as an intermediary within a distributed sensing system, where agency emerges through interactions between human, technological, and marine entities. In a subsequent sequence, the camera focuses on her silent operations along the shoreline as she deploys the network of hydrophones and cables across the rocky terrain. Her meticulous engagement with the technical equipment takes the form of a gracious dance as she moves, orchestrates, measures, calibrates, and listens. Hydrophones and cables compose a complex network of technical sensors spread across the shoreline's jagged surface, being prepared for submersion. The narrative tempo in the video is deliberately modulated through digital visual effects that dissolves frames and compress time. This cinematic technique emphasizes the time-intensive nature of the aquanaut's work as she tunes and co-directs the instruments in action. It also highlights how *medianatures* goes beyond spatial or material entanglements to encompass temporal ones. Parikka contrasts solid, object-like materialities with non-solid, processual media forms such as waves, frequencies, and electrical currents that unfold over time and often elude direct human perception. A low, ground-level camera angle places the viewer within the technological assemblage, as if seeing from the perspective of a hydrophone lying on the rocky shore, poised for deployment. This framing not only accentuates the technical and operational aspects of the project but also foregrounds the role of the hydrophones as technical mediators between the human and their environment. The hydrophones act, in Simondon's (2017) terms, as transductive agents—modulating relations between human and the environment through their participation in a distributed sensing milieu.

Another shot shows a close-up that delineates the boundary where the ocean interfaces with the land. The silent narration reappears, describing this region above the Arctic Circle as "insomniac" during the summer months, when continuous daylight disturbs the sonic particles through its influence on the water. The midnight sun profoundly affects the northern North Atlantic Ocean, subtly altering the thermal profile and stratification of its waters. These seasonal shifts in temperature create a dynamic environment where sound waves travel at varying speeds, bending and refracting through layers of differing densities (Affatati, Scaini, and Salon, 2021). This naturally occurring acoustic variability is increasingly destabilized by human-induced climate change, which alters temperature gradients and disrupts underwater communication and navigation systems. The rising acidity of ocean waters further modifies the absorption of sound, particularly diminishing the propagation of low-frequency waves (Madiligama and Zhang, 2023). Simultaneously, the influx of freshwater from melting ice and rivers alters salinity and thermal stratification, adding further instability to the sonic environment. These interrelated factors—thermal, chemical, and hydrological—shape the evolving underwater soundscape in which *Acoustic ocean* is situated. As the subtitles explain: "Only in the Arctic does everything surface."

The aquanaut's laborious engagement with the equipment underscores the thematic exploration of human interaction with marine ecosystems, mediated by sophisticated technical instruments. The next shot adopts the viewpoint of the hydrophones as they breach the ocean surface. A subsequent image presents a tableau of drifting marine organisms—possibly plankton or small cnidarians such as jellyfish or sea anemones. Their translucent forms catch the ambient light, while the ocean's low-frequency hum, punctuated by subtle variations, envelops the scene. The subtitles identify a sea butterfly as it comes into view, most likely *Limacina helicina*, a species that inhabits the High North Atlantic, including the waters around the Lofoten archipelago. These pelagic zooplanktonic sea snails are essential to the marine ecosystem, serving both as prey and as indicator species for environmental change. Acting as chemical recorders of rising acid levels, they serve—as the subtitles explain—as a canary in the coal mine, signaling the ocean's ecological vulnerability. Known as pteropods, these tiny organisms—measuring less than 1 cm—take on an imposing presence as the video crosses a scalar threshold. Their close-ups, now magnified to cover the full surface of the cinema screen, exemplify what DiCaglio (2021) describes as the systematic expansion of observable phenomena through scale, and resonate with Horton's (2021) argument that developing scalar literacy is vital to understanding how environmental and epistemological processes unfold across nested systems. This is not merely a visual effect, but an operative condition that configures the ocean as a mediated and relational space. As Bull and Miletic (2025) argue, scale is not merely a representational tool, but an active force that structures perception through technological and cartographic regimes, shaping what can be sensed and known. In *Acoustic ocean*, the transition from microscopic marine life to immersive projection does more than magnify an image—it draws the viewer into a reconfigured sensory hierarchy, one that foregrounds multispecies entanglements and alternative modalities of environmental sensing.

This scalar and sensory transformation in *Acoustic ocean* exemplifies what Dillon (2012) describes as a defining gesture of Indigenous futurism: the speculative reordering of epistemic hierarchies through aesthetic means. The use of remote sensing technologies is framed as a world-building gesture, one that centers nonhuman perspectives and interweaves Indigenous knowledge with posthuman techno-poetics. Within this framework, the interplay between human sensing, technological apparatuses, and marine life is shaped through the processes of amplification, attenuation, and filtration, underscoring Gabrys' (2016) argument that sensing technologies not only extend perception but actively reconfigure environmental relations. The hydrophones do not simply record, they participate in the conditions of sensing, filtering and producing the very relations they render perceptible. In turn, the aquanaut's gestures make these relational, multispecies dynamics sensible through practice, care, and mediation. This vision aligns with Sámi epistemologies, which reject binary divisions between human and nature, instead conceiving landscapes, waters, and nonhuman life as co-constitutive of human existence (Gilbert et al., 2023). Here, scale becomes a structuring rhythm, guiding how perception unfolds across species, sensors, and environments while shaping the kinds of ecological futures that can be imagined.

The magnified sea butterflies, whose delicate forms reflect the health of the ocean, embody this shared vulnerability. Serving as sensitive indicators of ocean health, their aragonite shells visibly deteriorate under the effects of acidification, registering the slow violence of altered chemistry across global waters. Their amplified presence on the large screen folds micro-scale life into the temporal and perceptual registers of planetary crisis, making visible the otherwise imperceptible processes that shape marine futures. When the camera returns to the shoreline, the aquanaut is in her tent, engaged in an analytical review of photographs depicting the deep-sea fauna previously observed. Equipped with navigational charts and a laptop computer, she meticulously examines a series of printed photographs that offer a tangible connection to the preceding underwater sequence. A subsequent transition returns to the marine depths, where the fauna are observed in their natural habitat. Accompanying the scene is a low, intensifying hum that crescendos into a near-ring, signaling a mounting urgency. At this moment, another subtitle emerges: "[t]heir absence will silence the submarine springs"—a line that echoes what Parikka (2015) describes as the recursive entanglement of media systems and ecological processes: the loss of species not only diminishes biodiversity but reconfigures the very grounds of perceptibility, calling attention to the deep-time materialities and infrastructures that condition both life and sensing.

At the heart of *Acoustic ocean* is the figure of the aquanaut, who navigates the seabed equipped with hydrophones and sensors, embodying the hybrid role of Indigenous scientist, artist, and mediator. Her presence underscores the mediated nature of human-ocean interaction, wherein knowledge production is not a process of extraction but a reciprocal exchange between human and nonhuman actors. In this sense, the aquanaut and her tools form an integrated network in which the human body, technological apparatus, and marine environment interact through the ongoing processes of modulation and mediation. Rather than merely recording data, these sensing infrastructures co-construct the ocean as an intelligible space, generating ecological knowledge through their material and infrastructural entanglements. In accordance with Parikka, this shows how media are not external to the environments they depict but are themselves ecological formations—medianatures—that participate in shaping the conditions of perception and life. The hydrophones do not simply capture pre-existing sounds; they render the ocean audible in new ways, actively configuring both its material presence and the human sensory relation to it. Sámi epistemologies similarly foreground such situated, land- and sea-based forms of knowing, where knowledge emerges through active participation in a relational field of human and nonhuman actors. In this view, sensing is an act of situated reciprocity and care.

The text overlays in *Acoustic ocean* weave between scientific descriptions—such as the principles of underwater sound propagation—and poetic reflections on the ocean's hidden agencies, enacting a speculative mediation that challenges extractive and colonial ways of sensing the sea. These overlays intermittently fade in over visuals of the aquanaut deploying hydrophones, traversing rocky shorelines, and submerging into underwater ecologies. Accompanied by a layered soundscape of cetacean clicks, distant rumbles, and crashing waves, the text not only provides information, but engages in speculative mediation, framing the ocean as a site of planetary interconnectedness. This speculative discourse is shaped by a key tension: the systematic erasure of Indigenous presence from both historical narratives and contemporary representations of marine environments across Lofoten, Vesterålen, and Senja. Echoing Indigenous futurist traditions (Dillon, 2012; Kite and Lewis, 2021), the work encapsulates Indigenous epistemologies not as mere folklore, but as future-oriented, technologically engaged world-building practices. Positioned at the intersection of ecological vulnerability and technological mediation, the aquanaut embodies a hybrid role which bridges dominant scientific frameworks and marginalized Indigenous knowledge systems. Her multispecies engagement highlights environmental understandings that go beyond extractivist approaches, emphasizing the ocean not merely as a resource but as an active medium embodying histories of climate change, species migration, and colonial displacement. By foregrounding the fragility of marine ecosystems, the work underscores that coexistence—across species and knowledge systems—depends on an embodied and reciprocal engagement with the ocean as a living, relational entity.

Indigenous Epistemologies in Artistic Research

In *Acoustic ocean*, knowledge emerges not only through Biemann's artistic practice but also through her collaboration with Sámi musician and activist, Sofia Jannok, who both acts in the film and informs its direction. In this way, the project exemplifies Henk Borgdorff's definition of artistic research, where new insights are generated through the intertwined artistic processes of making and inquiry (Borgdorff, 2012). Biemann's multisensory installation is not only the outcome of research but also an active method of inquiry, engaging with marine ecosystems, Sámi epistemologies, and technological mediation. In this context, the work enacts a relational form of knowledge production—one that aligns with and learns from Sámi understandings of knowledge as an ongoing and reciprocal exchange with the environment rather than a detached act of measurement. From a Sámi perspective, the act of listening to the ocean is mirrored in the slogans of many of the Indigenous environmental movements, such as "water is life," "land back," and the particular Sámi "Ellos eatnu" and "Gulahallat Eatnamiin/We Speak Earth" performed by Sámi artists who are also activists, such as Ella Marie Hætta Isaksen. The "Gulahallat Eatnamiin," which means "communicating with Earth," is a song about listening to the Earth which was performed by Sara Marielle Gaup Beaska in preparation for the COP 21 Climate Convention in Paris. This emphasis on listening as an active, relational practice not only aligns with Biemann's approach as an artistic researcher but also resonates with what Helmreich (2016) describes as "sounding the limits of life," where acoustic practices disclose more-than-human entanglements in planetary systems.

Ursula Biemann's *Acoustic ocean* (2018) also forms part of a broader artistic research discourse that rethinks environmental sensing as a relational and co-constitutive process. This approach aligns with recent shifts in media theory and Indigenous studies that foreground the role of mediation in shaping ecological relations and planetary imaginaries. Comparable explorations appear in the artistic works of Susan Schuppli, Emilija Škarnulytė, and Karrabing Film Collective, each engaging moving image practices to critically examine ecological crisis, extractivism, and more-than-human entanglements—though distinct approaches, ranging from sensor-based installations to narrative and place-based storytelling. Biemann's perspectives, furthermore, resonate with the concepts of stewardship and care for nature as theorized by Liv Østmo (the concept of jávredikšun). Gilbert et al. (2024) also refer to Sámi seascape caring practices as being vital to the continuitiy of sustainable nature harvest and fisheries in the Porsanger fjord in Finnmark. Whereas these caring practices are associated with small-scale harvest and fjord ecosystems, there is no reason why Sámi values and practices embedded in cultural continuity would cease to exist with a change of environment from fjord to open ocean fisheries, or from traditional to modern technology.

In the paper, "A cybernetic future for small-scale fisheries," Brattland (2014) emphasizes that socio-technical systems are no less "technical" just because they rely on traditional technology such as oars and sails. An important point is that the technological system consisting of the material employed in the fisheries up until the 1950s—the wooden boat, the landscape, the hand line, the practice of creating a navigational system by naming places on the water in the intersection between land lines, giving names that connected a name with the topology of the landscape, was a technology adapted to the historical and social context of a particular time. It is a technology that is still in use today, but alongside other technologies such as Facebook, GPS, mobile phones, and echo sounders that are applied to find the same spot that was previously found, only using another kind of technology.[6] Might it be possible to preserve the values and ecological principles of caring for nature despite changing technologies and environmental contexts? This vision is realized in the figure of the environmentalist Indigenous presence of Jannok in the film *Acoustic ocean*, as a hybrid between Beret Paulsdatter and G.O. Sars. In *Acoustic ocean*, the ocean is not positioned as a passive landscape or resource but as an active participant in a multispecies network—a site where human, nonhuman, and technological agencies meet to negotiate new ways of sensing and relating to the more-than-human world. This gesture, from extraction to cohabitation, is what makes works like *Acoustic ocean* not just representations of ecological crisis, but interventions showing how ecological futures could be imagined and perceived otherwise.

Conclusion

The erasure of Sea Sámi presence from the cultural memory of the Lofoten, Vesterålen, and Senja regions is central to an understanding of the significance of the aquanaut figure in *Acoustic ocean*. This character—a kind of "Indigenous scientist"—raises questions about how scientific tools and Indigenous knowledge systems might be brought into relation. The figure of the aquanaut, navigating between knowledge systems, highlights how sensing is not a passive act of data collection, but a relational practice that redefines human-ocean interactions. In this way, *Acoustic ocean* offers an alternative imaginary—one that reintroduces Sámi epistemologies into oceanic discourse, presenting them not as remnants of the past but as integral to future ecological engagements.

Biemann's *Acoustic ocean* (2018) thus exemplifies how artistic research can function as a critical site for reconfiguring ecological knowledge through sensory and technological mediation. Through a multisensory approach, the installation foregrounds the ocean as both a site of material vulnerability and an active medium of transmission, bringing together environmental science, Indigenous knowledge, and experimental moving image practices. By positioning the Sámi aquanaut as a mediator, the work disrupts dominant extractivist frameworks and challenges the historical erasure of Indigenous epistemologies in ocean governance. These concerns gain particular urgency in light of the ongoing environmental and political tensions surrounding the Lofoten, Vesterålen, and Senja archipelagos. In contrast to inland environmental protests where Sámi activists have taken prominent roles, the LoVeSe movement unfolded without Sámi representation—not by exclusion, but because the coastal Sámi presence had already been rendered invisible through generations of cultural erasure, leaving only traces in place names, stories, and memories.

At the same time, the installation resonates with the speculative strategies of Indigenous futurisms, where ecological futures are articulated through the practices of relation, resistance, and survivance. Jannok's aquanaut embodies this repositioning, suggesting that artistic research not only reflects environmental entanglements, but reconfigures the frameworks through which they are sensed, narrated, and acted upon. If, as Brattland (2014) argues, ecological continuity in Sámi contexts has historically been sustained through place-based practices embedded in socio-technical systems, *Acoustic ocean* speculates on how such principles might extend into oceanic contexts shaped by digital instrumentation and environmental precarity. By adopting this relational approach, *Acoustic ocean* can be seen to extend Gabrys' concept of sensing as a form of world-making, while also engaging Parikka's notion of medianatures, where media are understood not as neutral channels but as ecological formations in themselves. Together, these frameworks invite viewers to reconsider how multispecies interactions and co-constitutive engagements with technology might inform alternative ecological futures.

While resistance to resource extraction in the region has gained visibility through environmental activism, Indigenous perspectives on ocean governance, multispecies relations, and marine stewardship remain underrepresented in both artistic and scientific discourse. Further research is needed to address this absence—not only by acknowledging Sámi cultural and historical entanglements with the ocean, but by exploring how Indigenous epistemologies might inform future ecological and political decision-making. Artistic research has a crucial role to play in this regard, offering methods of listening, sensing, and imagining that operate outside dominant epistemic frameworks. Rather than offering a resolved narrative, *Acoustic ocean* invites speculative engagement with marine futures, drawing attention to how oceanic environments are mediated through entangled technological, ecological, and epistemic relations. Situated at the intersection of media, ecology, and Indigenous knowledge, Biemann's work intervenes not by offering answers, but by reconfiguring the terms through which oceanic futures can be sensed, imagined, and narrated.

Notes

1 This description refers to the installation of *Acoustic Ocean* at the 2018 Taipei Biennale. Presentation formats may vary in other exhibitions. Images of the work can be accessed at the artist's website: https://geobodies.org
2 e-flux. (2018, September 20). *The Atlantic Project: After the Future.* https://www.e-flux.com/announcements/211966/the-atlantic-projectafter-the-future/ The work premiered in Eco-Visionaries at the House of Electronic Arts in Basel, August 2018 Among many exhibitions it has been shown at the 2018 Taipei Biennale; Centre culturel suisse Paris, 2020; Indigenous Knowledge_Cosmological Fictions at MAMAC Nice, 2020; International Design Festival Beijing 2021.
3 Henry Buehler, *Norway's Identity Crisis: The Battle for Lofoten*, Arctic in Context, October 31, 2017, https://jsis.washington.edu/aic/2017/10/31/norways-identity-crisis-the-battle-for-lofoten/
4 NOAA Ocean Explorer, *Acoustics: SOFAR Channel*, 2001, https://oceanexplorer.noaa.gov/explorations/sound01/background/acoustics/media/sofar.html
5 The Lofoten archipelago is located above the Arctic Circle. It lies between 67 and 68 degrees north latitude in Norway, well within the Arctic region. This location results in phenomena such as the midnight sun in the summer and the polar night in the winter.
6 Indigenous navigation methods are a form of mapping as a social practice (Kitchin and Dodge, 2017) as discussed in the chapter by Asko, this volume.

References

Affatati, A., Scaini, C., and Salon, S. (2021). Ocean sound propagation in a changing climate: Global sound speed changes and identification of acoustic hotspots. *Earth's Future*, 9(3). doi:10.1029/2021EF002099.

Andaur, K., Gulowsen, T., Losnegård, S.H., Schjølset, S., and Riise, A.B. (2025, February 28). Nei til oljeboring i Lofoten [No to oil drilling in Lofoten]. *Nordnorsk Debatt*. https://www.nordnorskdebatt.no/nei-til-oljeboring-i-lofoten/o/5-124-352087.

Asdal, K. and Huse, T. (2023). *Nature-made economy: Cod, capital, and the great economization of the ocean*. MIT Press.

Biemann, U. (2018). *Acoustic ocean* [Video installation].

Borgdorff, H. (2012). *The conflict of the faculties: Perspectives on artistic research and academia*. Leiden University Press.

Brattland, C. (2014). A cybernetic future for small-scale fisheries. *Maritime Studies*, 13 (18). doi:10.1186/s40152-014-0018-1

Buehler, H. (2017, October 31). Norway's identity crisis: The battle for Lofoten. Arctic in context. Henry M. Jackson School of International Studies, University of Washington. https://jsis.washington.edu/aic/2017/10/31/norways-identity-crisis-the-battle-for-lofoten/.

Buhl-Mortensen, L., Buhl-Mortensen, P., Dolan, M., and Dannheim, J. (2018). Impacts of bottom trawling and litter on the seabed in Norwegian waters. *Frontiers in Marine Science*, 5, 42. doi:10.3389/fmars.2018.00042.

Bull, S.T. and Miletic, D. (2025). *Proxistant vision: Motion, navigation, scale*. MIT Press.

Demos, T. J. (2016). *Decolonizing nature: Contemporary art and the politics of ecology*. MIT Press.

DiCaglio, J. (2021). *Scale theory: A nondisciplinary inquiry*. University of Minnesota Press.

Dillon, G. (Ed.). (2012). *Walking the clouds: An anthology of Indigenous science fiction*. University of Arizona Press.

e-flux. (2018, September 20). The Atlantic Project: After the future. https://www.e-flux.com/announcements/211966/the-atlantic-projectafter-the-future/.

Fjellheim, E.M. (2023). Wind E on trial in Saepmie: Epistemic controversies and strategic ignorance in Norway's green energy transition. *Arctic Review on Law and Politics*, 14, 140–168.

Fossen, C. (2020, September 23). Norway allows widespread and destructive fishing in protected sea areas. *Science Norway*. https://www.sciencenorway.no/nature-conservation-oceans/norway-allows-widespread-and-destructive-fishing-in-protected-sea-areas/1843361.

Gabrys, J. (2016). *Program earth: Environmental sensing technology and the making of a computational planet*. University of Minnesota Press.

Gabrys, J. (2019). Sensors and sensing practices: Reworking experience across entities, environments, and technologies. *Science, Technology, & Human Values*, 44(5), 723–736. doi:10.1177/0162243919852673.

GCE Ocean. (n.d.). Love ocean observatory. https://www.gceocean.no/projects/gce/love-ocean-observatory/.

Gilbert, J., Brattland, C., de Maat, S., Kramm, M., and Pelizzon, A. (2023). Caring for nature: Exploring the concepts of stewardship in European philosophies, spiritual traditions, and laws. In J. García Ruales, K. Hovden, H. Kopnina, C.D. Robertson, and H. Schoukens (Eds.), *Rights of Nature in Europe: Encounters and Visions*. doi:10.4324/9781003318989-4.

Haraway, D.J. (2016). *Staying with the trouble: Making kin in the Chthulucene*. Duke University Press.

Hansen, M.B.N. (2015). *Feed-forward: On the future of twenty-first-century media*. University of Chicago Press.

Helmreich, S. (2016). *Sounding the limits of life: Essays in the anthropology of biology and beyond*. Princeton University Press.

Hildebrand, J.A. (2009). Anthropogenic and natural sources of ambient noise in the ocean. *Marine Ecology Progress Series*, 395, 5–20.

Horton, Z. (2021). *The Zoom: How everything moves together*. University of Chicago Press.

Kitchin, R., and Dodge, M. (2007). Rethinking maps. *Progress in Human Geography*, 31(3), 331–344. https://doi.org/10.1177/0309132507077082

Kite, S. and Lewis, J.E. (2021). Making kin with the machines. In H.A. Davis and A.M. Weheliye (Eds.), *Beyond the human: A feminist future for AI* (pp. 59–75). MIT Press.

Kristoffersen, B., Bridge, G., and Steinberg, P. (2021). Time for oil: Competing Petro-temporalities in Norway's Lofoten/Vesterålen/Senja Archipelago. In F. Polack and D. Farquharson (Eds.), *Cold water oil* (pp. 176–193). Routledge.

Madiligama, M. and Zhang, L. (2023). Climate-induced changes in temperature, salinity, and sound speed in the global ocean. *The Journal of the Acoustical Society of America*, 153(3), A26-A26. https://pubs.aip.org/asa/jasa/article/153/3_supplement/A26/2885437/Climate-induced-changes-in-temperature-salinity.

Marine Stewardship Council. (2021). What is bottom trawling?https://www.msc.org.

Neimanis, A. (2017). *Bodies of water: Posthuman feminist phenomenology*. Bloomsbury Academic.

Nielsen, A.R. (2024). *Samisk Fortid i Lofoten*. Orkana.

NOAA Ocean Explorer (2001). *Acoustics: SOFAR channel. NOAA*. https://oceanexplorer.noaa.gov/explorations/sound01/background/acoustics/media/sofar.html.

Northrop, J. (1974). SOFAR sound speeds and axis depths in the North and South Atlantic Ocean. *Journal of the Acoustical Society of America*, 55(2), 418. doi:10.1121/1.3437307.

Norwegian Ministry of Energy (2024). The government proposes to increase the allocation for resource and environmental mapping of seabed minerals. https://www.regjeringen.no.

Oceana. (2021). Bottom Trawling. https://usa.oceana.org.

OceanCare. (2023). Bottom Trawling: New Report Highlights Need for Action. *OceanCare*. https://www.oceancare.org/en/stories_and_news/report-trawling/.

Parikka, J. (2015). A geologic media theory. In J. Parikka, *A geology of media* (pp. 1–25). University of Minnesota Press.

Public consultation of the first licensing round for seabed minerals. (2024). Press release (26/06/2024–03/12/2024). https://www.regjeringen.no/en/aktuelt/public-consultation-of-the-first-licensing-round-for-seabed-minerals/id3047008/.

Raheja, M.H. (2010). *Reservation reelism: Redfacing, visual sovereignty, and representations of Native Americans in film*. University of Nebraska Press.

Reuters (2024, December 1). Norway to stop deep-sea mining, says party supporting minority government. https://www.reuters.com/world/europe/norway-stop-deep-sea-mining-says-party-supporting-minority-government-2024-12-01/.

Richardson, W.J., Greene, C.R., Malme, C.I., and Thomson, D.H. (1995). *Marine mammals and noise*. Academic Press.

Simondon, G. (2017). *On the mode of existence of technical objects* (C. Malaspina and J. Rogove, Trans.). Univocal Publishing. (Original work published 1958.)

Staberg, M. and Budalen, A. (2024, January 26). Flere unge vil lete etter olje utenfor Lofoten, Vesterålen og Senja. NRK. https://www.nrk.no/nordland/flere-unge-vil-lete-etter-olje-utenfor-lofoten_-vesteralen-og-senja-1.17355779.

Tsing, A.L. (2015). *The mushroom at the end of the world: On the possibility of life in capitalist ruins*. Princeton University Press.

US Geological Survey. (2016). What a drag: The global impact of bottom trawling. https://www.usgs.gov.

Whyte, K. (2017). Indigenous climate change studies: Indigenous futures, decolonizing the Anthropocene. *English Language Notes*, 55(1–2),153–162. doi:10.1215/00138282-55.1-2.153.

Wikipedia contributors (2023, July 17). *SOFAR channel*. In Wikipedia, The Free Encyclopedia. https://en.wikipedia.org/wiki/SOFAR_channel.

13 Through the Eyes of the Greenland Shark

Imag(in)ing the High North Atlantic

Susanne Østby Sæther

Introduction

Entering the darkened gallery space of Marie Kølbæk Iversen's video installation *Histories of Predation* (2022), one is met by three monumental screens displaying mesmerizing, abstract shapes in bright colors, interspersed with fields of black. The imagery appears in endless vertical streams. Punctuating the streams is the recurring shift between magenta and green color fields, demarcated with a horizontal line that imbues the imagery with a machinic rhythm. While the imagery may appear abstract, it is in fact ultra-concrete: It displays microscopic recordings of the thousands of stratified fibers that form inside the inner eye lenses of the Greenland shark throughout its long life (Figure 13.1).[1]

Also known as the Gurry Shark, the Gray Shark or in Greenlandic/Kalaallisut, as *qalussuaq*, the Greenland shark (Somniosus microcephalus) is the shark species that lives farthest north in the world, with its distribution stretching from Northeastern North America, Greenland, Iceland, Svalbard, across the White Sea, and, towards the south, the Faroe Islands, the British Isles, Norway, and Skagerrak. Its transit zones thereby delineate quite precisely the waters we in this volume have called the High North Atlantic (see Figure 1.1 in the introduction). Indeed, Danish marine biologist Julius Nielsen, a leading expert on the species and one of Kølbæk Iversen's collaborators in making the work, has suggested that a species name that more appropriately would indicate its area of distribution would be the "Northern North Atlantic deep-sea shark."[2]

The Greenland shark, which can grow up to 7 meters in length and inhabits depths ranging from 0 to 1200 meters, is also the longest-living vertebrate. Through radiocarbon dating of its eye lens nucleus, recent marine biological research from the University of Copenhagen, led by Nielsen, has shown that the shark may live to become as old as 272 to 512 years, with individuals reaching reproductive maturity at around 150 years. In the Greenland shark, the innermost eye lens proteins are formed before birth and preserved unchanged, storing chemical traces of the environment at that time. As the shark grows, new layers of crystalline proteins form around the

DOI: 10.4324/9781003478805-13

Figure 13.1 Installation photograph of Marie Kølbæk Iversen's video installation *Histories of Predation* (2022), from the exhibition *The Atlantic Ocean. Art, Myths, Science*, Henie Onstad Kunstsenter, April 26 to August 15, 2024

Source: Photo: Øystein Thorvaldsen/Henie Onstad Kunstsenter.

original inner lens. Each layer preserves the chemical composition of the shark's environment at the time it was formed, like rings in a tree. Through radiocarbon dating, Nielsen and his colleagues have calibrated and converted such biochemical traces in the shark's inner eye nucleus into age estimates.

Considering the longevity of the Greenland shark, the chemical traces recorded in its eye lens may originate from a temporal horizon that far exceeds a human lifetime. This extensive temporal scale allows Kølbæk Iversen to – through the eyes of the Greenland shark—"reflect on 500 years of colonial, imperial and environmental battles in the northern Atlantic region, where Danish influence was and is significant," as stated in a catalogue presentation on the work (Lepoutre Ravn, 2022, unpaginated). An individual of the species living (or caught) today can thus be conceived as a potential "eye witness" to the various predatory endeavors that have taken—and still are taking—place in the waters we here call the High North Atlantic (Kølbæk Iversen, 2022). More specifically, the shark may be conceived as what Michael Richardson (2024) has recently termed a "nonhuman witness" to these events, referring to the ability of nonhuman entities—from technologies and algorithms to ecological phenomena and animals—to register and record events or their traces at various degrees of independence from human agency (2024, p. 3).

In its decentering of the human and its acknowledgment of the mediating capacities harbored in elements and entities other than traditional media technologies, *Histories of Predation* resonates with the elemental turn that has marked media theory in the last decade. Crucial in this turn is the reconception of mediation as a process involving elemental forces, such as air, water, minerals, and organic materials, in shaping matter, life, and perception across temporal and material scales (Cubitt, 2016; Jue, 2020; Parikka, 2015; Peters, 2015; Starosielski, 2019). Within an elemental approach, the ocean is understood not merely as a backdrop, but as an active medium that conditions existence and relations (Peters, 2015, pp. 39–41). Entanglements between media and natural elements are foregrounded also by Jussi Parikka's through his concept of "Medianatures," emphasizing that media materially emerge from and remain embedded within geological and ecological contexts (Parikka, 2018). However, while Parikka primarily addresses minerals and inorganic matter, *Histories of Predation* extends this discourse to organic life forms. Kølbæk Iversen, in her work, construes both the ocean and the shark itself as mediating forces: the ocean circulates matter, energies, humans, and stories, whereas the shark's eye lens stores and transmits environmental traces across centuries.

In this chapter, I take Kølbæk Iversen's video installation as a point of departure for exploring how specific dimensions of the High North Atlantic as a physical environment, as well as a cultural and geopolitical formation, and (predatory) practices therein may be foregrounded when registered, stored, and "witnessed" through the eyes of the Greenland shark. What relationship does Marie Kølbæk Iversen's work establish between the microscopic imagery of the stratified fibers of the lens nuclei that fill the screen and the past and future scenarios and visions of oceanic worlds these images may evoke? Or, differently put: What is the relationship between the *imagery* and the *imaginaries* of and from the High North Atlantic that *Histories of Predation* conjures?

Pursuing these questions, I take three steps. First, I examine the *imagery* that constitutes the work, informed by Shannon Jackson's concept of "eco-aesthetics" (2024). In a recent essay, Jackson explores how media and video art practices have responded to ecological concerns over the past two decades (2024). Building on features identified by Jackson as key in this aesthetic, I discuss the source material and methods Kølbæk Iversen has employed to create the work's vivid imagery, as well as the work's spatial and temporal configuration. Second, I discuss a set of historical and contemporary *imaginaries* of the High North Atlantic, identifying the Greenland shark as a monstrous and liminal figure within these imaginaries. Finally, to untangle the intersection of imagery and imaginaries in the work, I draw on the age determination methodology of marine biologist Julius Nielsen (2017). Speculatively expanding on Nielsen's conception of the shark's eye lens as a *biogenic archive*, I discuss how, in *Histories of Predation*, the eye lens figures as a natural record of environmental conditions and biological activity in the area over centuries, recorded and preserved in living matter. Finally, I suggest that Kølbæk Iversen extends this recording capacity also to encompass predatory human practices and cultural imaginaries of the region.

Histories of Predation: The Work in Context

Histories of Predation is part of Kølbæk Iversen's artistic PhD and was co-commissioned by Henie Onstad Kunstsenter for the exhibition *The Atlantic Ocean: Art, Myths, Science* (2024), curated by Susanne Østby Sæther, Stefanie Hessler, and Knut Ljøgodt.[3] However, the installation was first presented as a solo exhibition at O—Overgaden (2022), a non-profit institution for contemporary art in Copenhagen, and subsequently at Statens Museum for Kunst's (SMK) satellite exhibition space in Thy in North Jutland in 2023.[4] At both institutions, the work consisted of three videos projected on custom-built, translucent screens. Arranged as freestanding elements in the exhibition space, the projection screens form a site-specific, spatial installation. Additionally, at Henie Onstad, three bronze-cast sculptures were added. Molded from the jaw, tail fin, and lower body of an individual shark in a 1:1 scale and suspended from the ceiling at different heights, these sculptural elements enhanced the shark's bodily presence in the work.

Through a constellation of works spanning four centuries, the exhibition *The Atlantic Ocean. Art, Myths, Science*, at Henie Onstad integrated aesthetic, scientific, and mythical perspectives. It particularly foregrounded the northern North Atlantic as a dynamic site of ecological transformation, extractivist endeavors, labor, and colonization and forced migration, engaging visitors in rethinking human–ocean relations in this region. As such, this exhibition and Kølbæk Iversen's work resonate with—and contribute to—the "oceanic turn" that has been evident in contemporary art and curatorial practices since approximately 2010. Art's "oceanic turn" reflects a heightened awareness of the oceans' critical role in ecology, colonial practices, and climate change, and of the increased impact of blue humanities scholarship and critical ocean studies (Hessler 2020; Syperek and Wade 2024). Marked by an interdisciplinary shift from anthropocentric frameworks to multispecies relationalities, this turn is also characterized by artistic and curatorial strategies that emphasize oceanic connectivity and fluidity. Exhibitions like *Streamlines: Oceans, Global Trade, and Migration* (2015–16), curated by Koyo Kouoh, *Tidalectics* (2017–18), curated by Stefanie Hessler, *Ocean Imaginaries* (2017), curated by Linda Williams, and the projection series *An Oceanic Feeling* (2018), curated by Erika Balsom, are other examples of this tendency.[5] Many of these exhibitions employ curatorial methodologies and frameworks that conceptually and/or thematically aim to echo the ocean's inherent rhythms and movements, examining oceans as spaces of cultural memory, political critique, ecological crisis, and collaborative possibility (Hessler, 2020; Syperek and Wade, 2024).

Kølbæk Iversen's work can also be seen as part of the "eco-aesthetics" Shannon Jackson recently identified: a range of ecologically responsive practices that have developed within video and media art over the past 20 years (2024). As Jackson notes, these practices revolve around recurring themes and technical strategies that explore historical and geographical networks and relations, which resist straightforward representation and challenge our imaginative capacities (p. 12). She emphasizes the extensive material agency involved in contemporary video art, where works techno-aesthetically are intertwined with their environments in multiple ways. This entwinement arises through various factors, including the production technologies used (such as cameras and software), the application of imaging or data-collecting techniques from other disciplines, and the installation of the work in the exhibition space. Through these forms of entwinement, the boundaries between artwork and environment often remain deliberately porous, unsettling distinctions between "artificial and natural, human and non-human, media and site, figure and ground"—along with "other comforting dyads" (2024, p. 12), as Jackson contends.

Jackson builds her argument on media philosopher Erich Hörl's (2017) conception of an unfolding "denaturalization" of ecology that started in the late 20th century, especially from the 1970s and onward with (second-order) cybernetics, systems theory, and ecological thought expanding beyond natural sciences into media, cognition, and social domains (pp. 3, 9). Ecology is thus shifted from its traditional association with nature to a "denaturalized" and generalized conceptual framework. Increasingly encompassing technological, informational, and social systems, ecology is becoming a key framework through which contemporary forms of power, governance, and mediation are understood (pp. 1–3). This shift, which Hörl terms a "general ecologization of thinking," dismantles the modern dichotomy between nature and technology, positioning ecology as a central paradigm of non-modern rationality. As such, the concept of ecology no longer merely describes environmental relations but becomes a model for grasping the deep entanglements of media, technics, and life itself (pp. 2–3). For Jackson, contemporary media art is pivotal in exploring this "denaturalized" ecology through extensive use of materials, imagery, and technologies culled from fields outside of art. This artistic strategy can also be identified in Kølbæk Iversen's work, as addressed throughout the following discussion.

Imagery: Configuring Confocal Microscopy

One of the artistic modes that Jackson identifies as central to the present "eco aesthetic" in media art is artists' use of technologies, images, and data sets derived from popular media and the natural sciences. Examples mentioned by Jackson include the use of algorithmic data collection and advanced visualization techniques (2024, p. 23). Kølbæk Iversen's employment of microscopic images of the shark's eye nuclei makes *Histories of Predation* a case in point.

At first encounter, the imagery of *Histories of Predation* appears abstract. Yet, as the work unfolds, its biological origin transpires: fibrous textures suggest organic tissue, while color gradients and the rhythmic scanning line indicate their basis in microscopic visualization technologies. To generate these microscopic images, Kølbæk Iversen collaborated with Jonathan Brewer, Professor and Director of the Danish Molecular Biomedical Imaging Center (DaMBIC) at the University of Southern Denmark. Together, they experimented with imaging two pairs of Greenland shark eyes, sourced from individuals caught as bycatch in commercial fisheries.[6] Obtaining the specimens, Kølbæk Iversen worked with marine biologist Julius Nielsen, who at the time was based at the Greenland Institute of Natural Resources (GINR) in Nuuk.[7] This interdisciplinary collaboration across art, marine biology, and biomedical imaging underpins the visual and conceptual framework of the work.

Central to producing the work's imagery was the artist's use of confocal laser scanning microscopy (CLSM), an optical imaging technique widely used in cell biology and neuroscience since the 1980s. While conventional microscopes capture light from all depths of a specimen and therefore often produce blurry images, CLSM uses a pinhole in front of the detector to block out-of-focus light, thereby enhancing resolution and contrast. By scanning the sample plane by plane, confocal microscopy assembles detailed three-dimensional reconstructions. It is thus particularly suited to visualizing biological stratifications, such as the Greenland shark's lens nucleus. Imaging at greater depths and with higher resolution than conventional microscopic techniques, Kølbæk Iversen's use of CLSM contributes to the distinctive visual qualities and material presence of the resulting images.

From 2020 to 2022, Kølbæk Iversen and Brewer conducted numerous work sessions where they experimented with different placements of the eye lenses on the sample holder and explored various microscopic settings and imaging modes. Their approach was exploratory and open-ended, shaped by an interplay of different "cultures of vision" (Galison and Jones, 1998) grounded in art and bioimaging, respectively. "We wanted to see what we could see," they stated in a conversation.[8] Two variables proved particularly decisive in their imaging experiments: depth and color, which together shaped the imagery in *Histories of Predation*.

Depth is a crucial dimension of the imagery, as it both structured the process of its production and serves as a temporal indicator. As mentioned, confocal microscopy generates images sequentially by scanning layer by layer through the specimen (Nwaneshiudu et al., 2012, p. 4). Utilizing this feature, in *Histories of Predation* each of the three projected image streams corresponds to a different depth of the shark's eye lens nucleus. Since deeper layers in the lens date further back in time, the images also serve as interfaces to different temporal strata, indicated in the imagery by variations in texture, density, and luminosity. In the final installation, depth thus shifts from a technical parameter to an aesthetic and temporal dimension, inviting viewers to slowly attune to the durational ecological timescales embedded in the work.

Color became another crucial variable in configuring the final imagery. In bioimaging, color coding is not fully standardized but chosen by the researcher based on the microscope's settings to highlight specific features of the sample.[9] Brewer is trained as a physicist and generally, in his bioimaging work, uses color to enhance scientific analysis and communicability. As an artist, Kølbæk Iversen was more invested in how color modulated the aesthetic and affective qualities of the imagery. The green and magenta/pink featured in the final work were chosen by Brewer from the microscope settings.[10] While the colors remained unaltered, their interlacing with the black parts of the imagery and the images' potential as scientific data is however radically altered and aestheticized when projected and blown up to the 300 x 300 cm of the screens.

The transition from biomedical imaging to video installation is further reinforced by the work's spatial configuration. For each iteration of the work, the three screens are carefully placed and arranged depending on the conditions of the space available. Custom-built to fit the aspect ratios of the image, the format of the screens was derived from the square format of the microscopic imagery.[11] Departing from the widescreen formats typical of cinema and television, the square format reinforces the imagery's grounding in scientific imaging. Kølbæk Iversen's choice to use transparent screens enables the imagery to be viewed from various angles, establishing ever-changing image constellations across the three screens. Through these elements, the installation compels viewers to navigate the exhibition space and heighten awareness of their embodied positioning in relation to the work. In this way, *Histories of Predation* foregrounds the interrelations between imagery, environment, and spectator, enacting what Jackson identifies as the active entwinement between media artworks and their immediate, material surroundings as well as with other disciplines and fields (2024, p. 23). By extension, Kølbæk Iversens's installation, along with many other contemporary "eco aesthetic" works, ultimately explores the relationship between humans and their environments.

Temporality is equally central to the work. Whereas in the imagery, depth indicates a stratified chronology in which deeper layers reach further back in time, the installation itself unfolds durationally. Each of the three video streams is constructed from microscopic stills, interlaced and animated with custom-made software. Edited into a 32-minute loop, the videos progress in synchronized cycles that emphasize cyclical rather than linear time and echo the scanning motion of the confocal microscope. This looping rhythm invites sustained attention, drawing viewers into the temporal structure of the imagery as it unfolds. The result is a meeting between two forms of temporality: the cyclical, durational time of the installation and the long biological and ecological timescale inscribed in the eye lens nuclei. As Jackson notes, such speculative temporal explorations are defining for the "eco aesthetic" in media art, as they provide perceptual and imaginative access to scales of duration that exceed human perception, yet that are constitutive of what has been termed the Anthropocene epoch (2024, pp. 22–23).

Ocean Imaginaries: Construing the Greenland Shark

During the opening and closing days of the exhibition *The Atlantic Ocean* at Henie Onstad Kunstsenter, Marie Kølbæk Iversen activated her installation through two performances. After offering the audience a toast of aquavit from a shell found on the shoreline outside the art center, she started to sing. Moving around between screens and sculptures, she performed revived folk songs from her West Jutland ancestry in local dialect, including *O'Tilli*, which recounts a mermaid abducted by a queen to serve as an oracle. Shared by the songs was the theme of suppressed identities and alternative histories of belonging, displacement, and resistance. Between each song, Kølbæk Iversen employed the bronze casts of the shark's body as percussion instruments, adding a ritual flair to the act. The performance thereby imbued the installation with "ocean imaginaries" of liminal beings and dissident "others" of the sea as well as of colonial and predatory attempts to subsume or silence such presences.

"Ocean imaginaries," "oceanic imaginaries," and similar terms frequently occur in the literature in the blue humanities field, drawing on the longer trajectory of the term "imaginary/imaginaries" in social sciences. Within blue

Figure 13.2 Documentation of Marie Kølbæk Iversen's performance *Donnimaar* (April 27, 2024), accompanying *Histories of Predation* (2022) in the exhibition *The Atlantic Ocean. Art, Myths, Science*, Henie Onstad Kunstsenter, April 26 to August 15, 2024

Source: Photo: Øystein Thorvaldsen/Henie Onstad Kunstsenter.

humanities, "ocean imaginaries" generally refers to how the ocean is conceptualized and represented in cultural and artistic contexts and the ocean's role in human history and mythology—more recently frequently intensified through concerns for environmental risks and climate change. *Ocean Imaginaries* was, for example, the title of an art exhibition at RMIT Gallery in Melbourne in 2017, focused "on some of the contradictions and conflicted feelings raised by how the ocean is imagined in an age of environmental risk."[12] Literary scholar Elizabeth Deloughrey has stated that "the ocean is now understood in terms of its agency, its anthropogenic pollution and acidity, and its interspecies ontologies—all of which suggest that climate change is shaping new oceanic imaginaries" (2017, p. 34). New materialist, feminist, and environmental humanities scholar Stacy Alaimo has noted that understanding ocean imaginaries is an important undertaking since humans tend to "imagine that the seas are imaginary rather than to undertake the scientific, cultural, and political work necessary to trace substantial interconnections between human discourses, human practices, and marine habitats" (Alaimo, 2012, p. 179, as cited in Braverman and Johnson, 2020, 8). Following Alaimo's argument, the notion of "ocean imaginaries" is particularly apt to Kølbæk Iversen's work, since it operates precisely at the intersection of art and science, human and non-human interconnections, and conjures a network of predatory behaviors across "human practices and marine habitats" (Alaimo, 2012, p. 179, as cited in Braverman and Johnson, 2020, 8).

Significantly, imaginaries associated more specifically with the distribution area of the Greenland shark—or the "Northern North Atlantic Deep Sea Shark", if one prefers—is presently undergoing profound changes. As Kirsten Thisted et. al. have argued, in a Northern European context, the designation "the North Atlantic" has variably been used to refer to a region that geopolitically is under construction—or reconstruction—consisting of Greenland, the Faroe Islands, Iceland, Norway, and Denmark in variable constellations (also including Orkney and Shetland Islands). "The New North Atlantic" is the designation Thisted and Gremaud have suggested for this variably defined area and its associated shifting imaginaries. Whereas "The New North Atlantic" designates a far more limited area than what we call the High North Atlantic, their arguments are equally applicable to the broader area addressed in this volume. With climate changes, including melting ice and warming oceans, the boundaries between the New/High North Atlantic and the Arctic are also renegotiated, with the "white Arctic turning blue" due to melting ice, and definitions of the Arctic moving from predominantly scientifically to politically grounded (Thisted and Gremaud, 2020, p. 12). An implication is that "old power relations between center and periphery are being destabilized" and power is moving north toward the Arctic. On the one hand, new, strategic partnerships and connections are being forged between previously marginalized societies and states (2020, p. 1). On the other hand, major state and non-state actors increasingly assert strategic and economic interests in the Arctic and High North Atlantic, seeking control over territories and resources in ways that risk undermining the political agency and rights of Indigenous and local populations in the area.

A vivid manifestation of shifting and contested power structures and geopolitical configurations in the High North Atlantic has historically been the depiction of sea monsters in maps, as evidenced by region maps from the 15th, 16th, and 17th centuries. Historian Erling Sandmo (2020) has demonstrated how the array of fantastical sea monsters illustrated in Olaus Magnus's pioneering map of the Nordic countries and surrounding seas, *Carta Marina*, printed in Venice in 1539, highlights the coexistence of diverse, often competing epistemological frameworks.[13] Monsters, such as sea serpents, whale-like "spouters," and a "sea pig," comprise a multifaceted reflection of power dynamics and human and animal predatory practices in the region. Whereas earlier, Medieval maps generally were not concerned with water, about half of the space of *Carta Marina* is, in contrast, devoted to northern seas, "filled with life and images," as Sandmo points out (2020, p. 240). Much of the sea area covered by the map is the Norwegian Sea, which geographically and oceanographically is considered part of the North Atlantic Ocean. As *Carta Marina* exemplifies, there is a genealogy of monstrous "ocean imaginaries" conveying shifting power relations in the High North Atlantic, to which *Histories of Predation* can be added.

A "monster with many hundreds of millions of years of evolution behind it, with potentially fatal poisons in its bloodstream and teeth like that of an oversized steel trap, only a lot more of them" (Strøksnes 2016/2018, p. 16). This is how Norwegian author Morten Strøksnes describes the Greenland shark in his best-selling documentary tale about his attempts to catch one in the Lofoten archipelago, translated to English as *Shark Drunk. The Art of Catching a Large Shark from a Tiny Rubber Dinghy in a Big Ocean* (2018).[14] Indeed, the Greenland shark may appear monstrous due to its notorious slowness, size, and unappealing appearance, often featuring a luminescent parasite protruding from its eyes, leading to near blindness. In contrast to cetaceans like blue whales, admired for their magnitude and songs, orcas, awed for their force and killer instinct, and dolphins, treasured for their kinesthetic and social intelligence, the Greenland shark is exceptionally uncharismatic in both appearance and behavior. It is illustrating that Inuit myths portray the shark in utilitarian and negative terms, contrasting it with revered marine animals. This negative depiction conveys the shark's perceived nuisance to fishing communities, as it frequently destroys fishing gear and has generally not been considered a source of food for humans (Nielsen, 2017, p. 16).

Moreover, as argued by Kølbæk Iversen in her dissertation accompanying *Histories of Predation*, in folklore and mythology within its distribution area, the Greenland shark is construed as a hybrid, liminal being through its feeding habits and local naming traditions. Being both an active predator—in fact, a top one in its area—and a scavenger, the shark feeds on living prey, including a variety of fishes, invertebrates, and other quarry, as well as dead carrion. For Kølbæk Iversen, these feeding habits and the shark's slow, deep-sea movements align the shark with cycles of decomposition and renewal, biologically and mythologically (Iversen, 2022, p. 199). She extends this to regional folklore and mythology by associating the shark with merpeople in Nordic folkloric traditions and linguistics. As Kølbæk Iversen points out, across the Nordic-Germanic languages, the Greenland shark has traditionally been designated as a "merperson," a mythic figure that blends human and aquatic traits, embodying fluid boundaries between species and realms. As an interspecies being that crosses human and nonhuman worlds, life and death, the shark serves as a mediator between oceanic depths and human imaginaries. Linguistically, the Greenland shark is referred to as "merman"—"hákarl"/"havkallur"/"havkal"—in Icelandic, Faroese, and Danish, and in Norwegian and Swedish as "merwoman"—"håkjerringa"/"håkjäring." Through these designations, the shark can also be construed as a non-binary figure that destabilizes conventional gender dichotomies (Nielsen and Kølbæk Iversen, catalogue 2022, unpaginated).

Building on these geographic and linguistic trajectories, Kølbæk Iversen outlines a set of mythological axes of the High North Atlantic Ocean. Drawing on folklore from her East Atlantic home region in Western Jutland, her performances feature merpeople and mythic beings as dissident Others in relation to the Danish state and its authorities. The bronze casts of the shark's body parts provide a looming physical presence in the installation. With the shark's open jaw with "teeth like that of an oversized steel trap" (Strøksnes, 2018, p. 16), hung just at head height as if it could snap at a visitor at any moment, the sculptures add to the rich monstrous imaginaries of *Carta Marina*. Hence, in *Histories of Predation*, the Greenland shark emerges both as a geographical demarcator of the High North Atlantic and as a pivotal species in the "ocean imaginaries" of its waters.

Witnessing the Biogenic Archive: From Imagery to Imaginaries

How, then, are the predatory and liminal imaginaries of the High North Atlantic connected to the microscopic imagery of the stratified, inner eye nucleus of Greenland shark, previously outlined? Marine biologist Julius Nielsen's research provides a key entry point into this intersection, as his work on the Greenland shark's longevity rests on how biological materials function as *archives* of environmental conditions in the species' distribution area.

To study the Greenland sharks' lifespan, Julius Nielsen has cross-combined techniques from other studies, applying radiocarbon dating to the eye lens nucleus. While similar combinations existed in other studies of vertebrates, they had not previously been used for marine animals of unknown age (Nielsen, 2017, p. 34). Nielsen, in his study, points to the function of the shark's inner eye nucleus as a *biogenic archive* (Nielsen, 2017, p. 34): organic materials that preserve biological information about past environmental conditions. The concept of biogenic archives is used across disciplines like geology, marine sciences, archaeology, as well as paleoclimatology and paleoecology. In these disciplines, materials such as tree rings, corals, shells, and pollen are analyzed as natural records of past environmental conditions and ecological change (Bradley, 2015; Smol, 2009). While proteins in the inner eye nucleus of the Greenland shark preserve radiocarbon markers of the environment at its embryonic stage, the successive layers forming around it throughout its life register later environmental signatures in their chemistry. Signatures may reflect diet, ocean chemistry, and broader ecological shifts, such as the "bomb pulse," a sharp rise in radiocarbon levels caused by nuclear weapons testing starting in 1945 (Nielsen, 2017, pp. 35–36, 46–47, 49–50). It is in this sense that the eye lens may serve as a biogenic archive. Such archives are critical for reconstructing long-term climate and ecosystem developments in the absence of human records.

Biogenic archives can be considered a subcategory of *natural archives*, such as ice cores, peat bogs, and ocean sediments, which record environmental changes over time. While conceptions of natural archives have been in use for over 200 years (Toepfer, 2019, p. 22), historians are increasingly integrating such sources into historical analysis, as Zachary Nowak observes (2022 [2019], p. 106). However, as Georg Toepfer reminds us, natural archives are not readily legible but contain proxy data, which require considerable human interpretation and technological mediation to function as archives (2019, p. 29). Cultural archives are, by contrast, often defined as systems of intentional classification and storage, structured by administrative power and feedback loops that shape future governance, as argued for instance by Wolfgang Ernst (2004; Toepfer, 2019, p. 30). While natural traces lack this original intentionality, they become archival in Ernst's sense through human activation—through ordering, extraction, processing, and their use in political or scientific arguments (Toepfer, 2019, pp. 30–31). In the Anthropocene, Nowak argues, the division between natural and human history becomes untenable, which makes the activation of natural archives a central historiographic practice (2020, pp. 21–22).

As a biogenic archive, the shark's eye lens may offer a glimpse into historical oceanic conditions, as Nielsen highlights (2017, pp. 54–55). However, in *Histories of Predation*, this glimpse is not simply about retrieving past environmental data; it also speculatively *testifies* to centuries of changing ocean conditions and predatory endeavors in the High North Atlantic, from pre-industrial ocean chemistry to the Anthropocene's radiocarbon bomb pulse and to markers of warming ocean temperatures, intensified extraction and shipping, and the colonial legacies in these Northern waters. Hence, the Greenland shark is construed as what Richardson (2024) terms a *nonhuman witness* to these events. In contrast to human testimony, nonhuman witnessing "exceeds speech and language" and includes material traces and affective endurance that register ecological and political transformations beyond human perception (2024, pp. 28–29). Nonhuman witnessing in this sense involves attending to materials and beings whose temporalities surpass our own, and whose bodies register historical and ecological violence in slow, sedimented ways (Richardson and Zolkos, 2022, pp. 3–5).

In discussing "witnessing ecologies," Richardson emphasizes the importance of scale: "Scale helps anchor perception in worlds that extend beyond the perceptual reach of the human sensorium; it enables one to conceive of entities far bigger or smaller, say, than can be contained within the human visual field" (2024, p. 136). This dynamic is evident in *Histories of Predation*. Through the extreme magnifying capacity and superior depth clarity of confocal microscopy, Kølbæk Iversen renders the eye nucleus as a biological archive perceptible, yet not thereby necessarily legible: Its fibrous strata are transformed into a vividly colored, extended temporal and immersive spatial experience, translating layers of lived environmental time into machinic rhythm and color. Here, as in ecological witnessing more generally, according to Richardson, scale shifts from being a tool of measurement to one of *address*, staging relations that immerse and implicate the viewer (2024, p. 137). Following this argument, *Histories of Predation* draws us into sensory, embodied engagements with scales that otherwise exceed our perceptual reach, binding human and nonhuman actors across scalar distances (2024, p. 137). Ultimately, then, the installation configures relations between media and milieu and in Jackson's terms, entwines the work with its outside in multiple ways (2024).

Through the Eyes of the Greenland Shark

By zooming in on the inner eye nucleus of the Greenland shark, *Histories of Predation* invites us to simultaneously zoom out toward the vast ecological and geopolitical forces that have shaped the High North Atlantic. Microscopic fibers of the shark's eye become a speculative lens that foregrounds developments, practices, and histories in and of the High North Atlantic that otherwise remain hard to perceive, yet are crucial in determining the conditions of our present and futures. This double movement demonstrates medial and conceptual strategies through which contemporary video art configures our access to oceanic histories, practices, and relations.

Through framing *Histories of predation* through Jackson's (2022) perspective on eco-aesthetics, the chapter demonstrates how Kølbæk Iversen transposes confocal microscopic imagery into an artistic register, establishing embodied, multisensory experiences of extended temporalities and materially inscribed ecological change. In situating the microscopic imagery of the shark's eye lens within the frameworks of ocean imaginaries and the biogenic archive, the chapter further indicates how biological material itself mediates some of the ecological changes, predatory practices, and extractive logics that have shaped the High North Atlantic over centuries. Underlying this analytical move is the contention that ocean imaginaries are not peripheral embellishments but central to how such practices are culturally registered and geopolitically contested, a point emphasized by Alaimo and other key contributors to the blue humanities (Alaimo, 2019; Braverman and Johnson, 2020).

Kølbæk Iversen's work can be argued to partake in what Hörl calls the denaturalization of ecology (2017), discussing how ecological relations are technologized and deeply entangled with various extractive logics. However, where Hörl reads this development primarily as a symptom of capitalist capture, Jackson emphasizes how eco-aesthetic practices can open artistic and political potentials within it. Building on Jackson, this chapter asserts that Kølbæk Iversen's work does more than diagnose: it suggests new modes of perceiving, experiencing, and imagining our complex, interlaced, and technologically mediated environments in and beyond the ocean. In materializing the links between images and imaginaries, the work offers a reflection on human entanglement with oceanic worlds and the enduring nonhuman presences that embody and silently witness our shared ecological trajectories. *Histories of Predation* thus reconfigures how we apprehend the High North Atlantic, inviting us to encounter and imagine these waters through the eyes of the Greenland shark.

Notes

1 *Histories of Predation* is the outcome of Kølbæk Iversen's artistic PhD and the result of a longstanding project of artistic research.
2 Conversation published in exhibition catalogue, O—Overgaden 2022.
3 The chapter author, Sæther, was the institutional curator for this exhibition and engaged in a longstanding dialogue with Kølbæk Iversen in advance of the exhibition opening, which shaped the work's iteration at Henie Onstad.
4 The exhibition period at O—Overgaden was August 27 to October 23, 2022, and at SMK Thy it was June 24 to August 27, 2023. While the work was scheduled to premiere at Henie Onstad as part of the exhibition *The Atlantic Ocean* in 2020 this was rescheduled due to the museum's closing during the COVID-19 pandemic and the consequences of the ensuing lockdown and postponement of planned exhibitions.
5 *Streamlines: Oceans, Global Trade, and Migration* was presented at Deichtorhallen Hamburg, while *Tidalectics* was presented at Thyssen-Bornemisza Art Contemporary (TBA21) in Vienna.
6 The shark is a common bycatch in the commercial fishery across the North Atlantic in countries such as, e.g., Greenland, Iceland, the Faroe Isle, Norway, and Canada. See Julius Nielsen and Marie Kølbæk Iversen (Katalog, 2022, pp. 10–11. https://overgaden.org/en/publications/rovhistorier).
7 In sourcing the eyes, the artist traveled to Nuuk to participate in dissecting an individual shark.
8 Teams conversation with Kølbæk Iversen and Brewer, October 15, 2023.
9 Conversation with Brewer, January 8, 2024.
10 The green color in imagery is based on autofluorescence in the sample itself, and the magenta/pink fields are based on scattered laser light, highlighting different features of the sample.
11 In confocal microscopy, image dimensions are user-defined but frequently rendered in square proportions.
12 https://rmitgallery.com/exhibitions/ocean-imaginaries/
13 Olaus' map covers the Baltic Sea, the North Sea, the Barents Sea, the Gulf of Bothnia, and the Norwegian Sea. The latter is the map's focus area and is geographically and oceanographically considered part of the Northern Atlantic Ocean.
14 The original title is *Havboka, eller Kunsten å fange en kjempehai fra en gummibåt på et stort hav gjennom fire årstider* (2016).

References

Alaimo, S. (2012). Dispersing disaster: The Deepwater Horizon, ocean conservation, and the immateriality of aliens. *American Quarterly*, 64(1), 101–123.

Alaimo, S. (2019). Introduction: Science studies and the blue humanities. *Configurations*, 27(4), 429–432.

Bradley, R.S. (2015). Paleoclimatic records from natural archives. In H.D. Holland and K.K. Turekian (Eds.), *Treatise on geochemistry* (2nd ed., Vol. 4, pp. 345–370). Elsevier.

Braverman, I., and Johnson, E.R. (Eds.) (2020). *Blue legalities: The life and laws of the sea*. Duke University Press.

Cubitt, S. (2016). *Finite media: Environmental implications of digital technologies*. Duke University Press.

Davies, S. (2017). *Ocean imaginaries: The ocean's eyes* [Exhibition catalogue]. RMIT Gallery.

DeLoughrey, E. (2017). Submarine futures of the Anthropocene. *Comparative Literature*, 69(1), 32–44.

Ernst, W. (2004). The archive as metaphor: From archival space to archival time. Open, 7, 46–52. https://s3.amazonaws.com/arena-attachments/542865/8b32821fe0174156942ede0cf145d55c.pdf.

Frank, S.K. and Jakobsen, K.A. (Eds.) (2019). *Arctic archives* (1st ed.). transcript Verlag. https://www.perlego.com/book/1527754.

Galison, P., and Jones, C.A. (Eds.) (1998). *Picturing science, producing art* (1st ed.). Routledge. doi:10.4324/9780203699942.

Hessler, S. (2020). Tidalectic curating. *Journal of Curatorial Studies*, 9(2), 248–270.

Hörl, E. (Ed.) (2017). *General ecology: The new ecological paradigm*. Bloomsbury Academic.

Jackson, S. (2024). Økologisk tid i tidsbaseret mediekunst. *Kunsten Som Forum På Tryk*, (14), 1–55. doi:10.7146/ksf.vi14.157089.

Jue, M. (2020). *Wild blue media: Thinking through seawater*. Duke University Press.

Kølbæk Iversen, M. (2022). *Neo-worlds: Transformative agency through fright, rite, and myth* (Doctoral dissertation). Oslo National Academy of the Arts & Aarhus University.

Lepoutre Ravn (2022). Rovhistorier. https://overgaden.org/en/publications/rovhistorier.

Nielsen, J. (2017). *The Greenland Shark (Somniosus microcephalus): Diet, tracking and radiocarbon age estimates reveal the world's oldest vertebrate* (PhD thesis). University of Copenhagen.

Nowak, Z. (2020). Theorising the natural archive. *Environment and History*, 26(3), 455–475. doi:10.3197/096734019X15740974883852.

Nwaneshiudu, A., Kuschal, C., Sakamoto, F.H., Anderson, R.R., Schwarzenberger, K., and Young, R.C. (2012). Introduction to confocal microscopy. *Journal of Investigative Dermatology*, 132(12), 1–5.

Parikka, J. (2012). New materialism as media theory: Medianatures and dirty matter. *Communication and Critical/Cultural Studies*, 9(1), 95–100.

Parikka, J. (2015). *A geology of media* (Vol. 46). University of Minnesota Press.

Parikka, J. (2018). Medianatures. *Zeitschrift für Medien- und Kulturforschung*, 9(1), 103–106.

Peters, J.D. (2015). *The marvelous clouds: Toward a philosophy of elemental media*. University of Chicago Press.

Richardson, M. (2024). *Nonhuman witnessing: War, data, and ecology after the end of the world.* Duke University Press.

Richardson, M. and Zolkos, M. (2022). Witnessing after the human. *Angelaki: Journal of the Theoretical Humanities*, 27(2), 3–16. doi:10.1080/0969725X.2022.2046355.

Richardson, M. and Zolkos, M. (2023). Witnessing the Anthropocene. *Angelaki: Journal of the Theoretical Humanities*, 28(4), 3–12. doi:10.1080/0969725X.2023.2233792.

Samuel, N. (2013). Images as tools. On visual epistemic practices in the biological sciences. *Studies in History and Philosophy of Science Part C: Studies in History and Philosophy of Biological and Biomedical Sciences*, 44(2), 225–236.

Sandmo, E. (2020). Dwellers of the waves: Sea monsters, classical history, and religion in Olaus Magnus's Carta Marina. *Norsk Geografisk Tidsskrift—Norwegian Journal of Geography*, 74(4), 237–249. doi:10.1080/00291951.2020.1810114.

Smol, J.P. (2009). *Pollution of lakes and rivers: A paleoenvironmental perspective.* John Wiley & Sons.

Starosielski, N. (2019). The elements of media studies. *Media+ Environment*, 1(1).

Strøksnes, M. (2016). *Havboka, eller Kunsten å fange en kjempehai fra en gummibåt på et stort hav gjennom fire årstider.* Oktober.

Strøksnes, M. (2018). *The art of catching a large shark from a tiny rubber dinghy in a big ocean.* Vintage

Syperek, P., and Wade, S. (2024). Oceanic curating. *Environmental Humanities*, 16(3), 680–691.

Thisted, K., and Gremaud, A.S.N. (2020). Envisioning the North Atlantic: Current narratives and official discourses. In Thisted, K., and Gremaud, A.S.N. (Eds.), *Denmark and the New North Atlantic: Narratives and Memories in a Former Empire*, 1, 9–74.

Toepfer, G. (2019). On similarities and differences between cultural and natural archives. In S.K. Frank and K.A. Jakobsen (Eds.), *Arctic archives: Ice, memory and entropy* (1st ed., pp. 21–36). transcript Verlag. http://www.jstor.org/stable/j.ctv371cfjt.5.

14 Cold-Water Coralations

On Petroculture and Mediation in the High North Atlantic

Melody Jue

An iconic sense of corals brings to mind warmth, the tropics, and color. Consider how the color- matching and design company Pantone promoted "Living Coral" as its color of the year in 2019, an orangish-pink hue that is "vivifying and effervescent," and "evocative of how coral reefs provide shelter to a diverse kaleidoscope of color."[1] This same rainbow palette has been used by media practitioners to exoticize corals. In *Coral Empire* (2019), Ann Elias cautions that in the early 20th century, "Photographers and cinematographers saw an opportunity to generate on film a coral orientalism," treating coral as a colorful and exotic other to promote a homogenous view of the tropics.[2] In biology, coral can be solitary (single) or colonial (made of many polyps). Globally distributed corals are cosmopolitan, and coral tourism sometimes blurs the specificities of place into a generic tropicality.[3] Even the United States' National Ocean and Atmospheric Association (NOAA) introduces corals by acknowledging that "most people think about clear, warm tropical seas and reefs filled with colorful fish."[4] This iconic sense of coral exhibits a warm-water bias that materializes everywhere from environmental activism to documentary media, marine science to the ocean humanities.[5]

Yet not all corals are tropical. While all corals are marine invertebrates with small stinging tentacles, related to jellyfish and sea anemones in the phylum Cnidaria, they come in many varieties: some are soft bodied, some have no symbiotic algae, and some live in cold, deep waters.[6] For example, the coral *Lophelia pertusa* (Figure 14.1)—more recently known as *Desmophyllum pertusum*—presents moments where common correlations, such as corals with color, warmth, tropicality, and symbiosis, break down.[7] *L. pertusa* is a bright white or orangish coral that thrives in depths, temperatures, and sometimes latitudes far away from what is normally associated with iconic coral. One of the largest reefs of *L. pertusa* in the world is Røst Reef, off the coast of Norway. *L. pertusa* can also be found in the Caribbean and west coast of Africa in deep-water environments—where depth provides a source of cold water, even in environments that might be considered tropical.

In this chapter, I examine how the connections between *L. pertusa* and forms of underwater mediation differ from those of tropical corals living in shallow-water environments. Choosing a non-tropical coral as a focus allows

DOI: 10.4324/9781003478805-14

Figure 14.1 A close-up of *Lophelia pertusa*
Source: Image courtesy of NOAA Ocean Exploration, Lophelia II 2009: Deepwater Coral Expedition: Reefs, Rigs, and Wrecks.

me to ask: how does this change in focal organism and environment shift our sense of coral's geopolitical and media-specific entanglements? Paying attention to the environmental conditions of mediation, and our concepts, exemplifies what I call milieu-specific analysis, a critical attention to how a concept holds up against specific environmental conditions and embodiments. As I argue in *Wild Blue Media* (2020), concepts take on different valences depending on the conditions of experience and perception within a specific milieu. So too with the concept of "coralations," a portmanteau that draws together coral and correlations, or coral and relations.

Coral-ations is a pun that has independently arisen across a variety of artistic and conservation contexts.[8] The nonprofit organization CORALations was founded in 1995 by Mary Ann Lucking and Orlando Peraza, and focuses on coral conservation in Puerto Rico and the wider Caribbean.[9] Artist Nicholas Magnan channels the pun in his film *Core Coralations* (2022), which dwells with an archive of coral core samples at the Australian Institute of Marine Science's Coral Core facility and the spectre of impending coral death.[10] In her short essay "Coralations: Returning to Breath" (2022), legal scholar and ethnographer Irus Braverman notes the important correlations between coral reefs and climate—of corals as indexical witnesses to past climates, and projections that corals will not survive future ocean warming events. She elaborates the portmanteau "coralations" as a way of thinking through the intimacies of corals as multispecies assemblages, co-saturating with law. For Braverman, the pun on coralations is about the specificity of coral relations—what corals are associated with (such as symbiotic algae) as well as what scientists and publics associate with or correlate with corals. Braverman writes that the idea of coralations came to her, "when contemplating the vibrant interrelations among various coral parts as well as between corals and other living beings."[11] Coralations, here, highlight conditions of interdependency.

Yet cold-water corals like *L. pertusa* pull into consideration different webs of ecological, cultural, and geopolitical coralations than their warm-water relations. Several associations common with tropical corals immediately go away: the sense of a rainbow reef, the ability to photograph corals through clear water, the visual allure of tourism, easy accessibility via snorkeling. Corals are often revered as icons of symbiosis, yet many other cold-water corals are azooxanthellate (without algae), and thus are not dependent upon the sun for metabolic functioning.[12] Somewhat ironically, despite living in dark waters, the name *Lophelia* comes from "lophos" (tuft of) and "helios" (sun). This follows from its bright white color and the star-like shape of each of its grasping polyps. One textbook notably proclaims, "*Lophelia* is the coral that breaks all the rules."[13]

I explore the implications of "*Lophelia* is the coral that breaks all the rules" for conceptualizing the media of the High North Atlantic—the organizing geography of this edited collection of essays.[14] Which forms of coralation are specific to *L. pertusa,* and what does this tell us about the biogeographic relationality of the High North Atlantic? How does *L. pertusa* align with or break warm-water coralations—for example, between corals and reef-formation, corals and symbiosis, corals and light/visuality? And what, especially, to make of its correlation with many sites of oil extraction?

Here I use the classification "cold-water corals" following Roberts et al. as the widely accepted term. *L. pertusa* was sometimes called a "deep-water coral" in older terminology. But while it can be found thousands of feet below the ocean, it has also been found in relatively shallow environments of 120 feet in the Trondheim Fjord (Trondsheimfjorden), which also possesses other supportive conditions: currents, food availability, and aragonite saturation, as well as seafloor composition and topography.[15] Because not all cold-water corals live in deep water, Roberts et al. argue that "cold" seems the more compelling umbrella term, quoting Thomas Wayland Vaughan: "It is temperature rather than depth that controls the bathymetric distribution of all corals except the shallow-water forms."[16] I retain the descriptor "cold" because of the way that it signals a difference from the iconic sense of corals in global environmentalism.

Part of the reason that cold-water corals have received far less attention than their warm-water relations has to do with the epistemological problem of access, and the political economy of oceanographic science. There are simply more scientists who study tropical corals than cold-water corals. This is perhaps because of the difficulty of observation—including the cost of research vessels, sonar, ROVs, and other media technologies to locate, observe, and collect specimens. It is far easier to access many tropical reef sites because of their shallower depths (closer proximity to light), which can be done easily from coasts and with smaller watercraft, or through scuba diving. Most cold-water corals thrive below recreational diving limits (120 feet), and benefit from remote observational technologies and ships that can weather more turbulent offshore conditions. The way that scientists and others learn about coralations—what correlates with corals, or about coral relations—is contingent upon the possibilities and processes of technical mediation.

The expense of studying cold-water corals, and other deep-sea organisms for that matter, has consequences for how knowledge is produced and disseminated. For example, the Allen Coral Atlas purports to "map the world's coral reefs," yet includes data on tropical corals alone.[17] Many other maps of global coral reefs follow suit, focusing on warm-water regions under the auspice of the global; comparatively few maps offer geographical comparisons with the locations of cold-water corals. Part of this may have to do with semantics: the Cambridge Dictionary defines a reef as a "a line of rocks or sand just above or just below the surface of the sea, often dangerous to ships," an old definition from sailing that excludes coral "reefs" that are too deep to cause shipwrecks.[18] Yet demographically, Roberts et al. write: "corals are primarily deep, cold-water inhabitants" with 65% of known species occurring in water below 150 feet. Anecdotally, one deep sea biologist with whom I spoke noted that warm-water coral biologists regularly peer review cold-water coral research—but said that he had never been asked to peer review any of their work.[19] It is perhaps unsurprising, then, that the humanities have followed suit, with book-length studies of corals almost exclusively focusing on warm-water stony (scleractinian) corals.[20]

In this chapter I examine two ways in which the coralations of *L. pertusa* differ from those of iconic, tropical corals. First, the technical mediations of sensing *L. pertusa* rely on sonar and connections to sound rather than visual media and analogies to cameras, as is more common with tropical corals in shallow, glassy water. The second coralation I examine is between *L. pertusa* and petroleum in the specific geopolitical context of the High North Atlantic. While tropical corals are not widely known for coinciding with petroleum infrastructure, *L. pertusa* is. Some of the first *L. pertusa* reefs were discovered as part of oil drilling surveys, and later, it was discovered to be growing on oil platforms. What *L. pertusa* models, then, is one way of contesting a normative view of tropical coralations that shape a common understanding of coral's iconicity. Following *L. pertusa* in the milieu-specific context of the High North Atlantic leads to its own set of coralations and relations to technical media—in particular, to sound and the infrastructures of oil extraction.

Broken Glass, Broken Photographic Analogies

I begin with a story recounted in the textbook *Deep-water Coral Reefs: Unique Biodiversity Hot-spots* (2008), written by engineer and marine geologist Martin Hovland—an employee of the Norwegian multinational energy company, Statoil (rebranded as Equinor in 2018).[21] One morning in July 1982, Hovland was mapping the seafloor with the vessel *Master Surveyor* for a future oil pipeline from the Askeladden field in the Barents Sea to Lyngenfjord, Norway. The sidescan sonar—a technology for producing visual maps of the seafloor from sound pulses and their reflections—identified a mysteriously cone-shaped structure, hundreds of meters below. To Hovland, the cone in the sonar map seemed highly unnatural: 15 meters tall and 50 meters wide. He wondered, "Was this, perhaps, a human-made submarine listening device, of the Cold War?"[22] The crew then deployed a "gravity corer" to assess the anomaly:

> Soon, the corer hovered at 260 metres depth, right above the top of the unidentified strange cone. Suddenly the corer was dropped until the wire went slack. As it was winched in, excitement rose. Most of us expected just another sticky clay core. *But we had luck and a sigh of astonishment went round as several pieces of white coral bits fell tinkling onto the steel deck, it sounded like bits of glass falling* [my emphasis].[23]

Norwegian waters were, after all, not the place that most people expected to find massive coral reefs. The coral that they encountered turned out to be *L. pertusa*, a "strange 'animal' with white 'eyes'" (Hovland 2008, 18). These eyes were polyps, the individual animals that collectively form the coral colony. What Hovland describes was not the first discovery of *L. pertusa* corals, nor the first encounter between the petroleum industry and seafloor ecologies, although it was the first time (his claim in *Deep-water Coral Reefs: Unique Biodiversity Hot-Spots*) that *L. pertusa* reefs were documented by the "modern imagery" of film and video via ROV deployment, which he reproduces in the book.[24] Seeing reefs matters: in their analysis of the Venus web portal offering real-time visual monitoring of reefs of *L. pertusa* off the coast of northern Norway, Elena Parmiggiani and Eric Monteiro examine how the "adoption of specific digital technologies in the monitoring infrastructure reshuffles power/knowledge and order relations within environmental risk assessment in the oil and gas sector," especially concerning the illusion of visual transparency from oil companies, whose financial interest in sub-sea monitoring would presumably be aimed towards future extraction.[25]

Yet what is remarkable in Hovland's passage is its description of sound. Not only does the coral poetically fall "tinkling onto the steel deck" like "bits of glass"; sound was an important epistemic mode for finding the coral in the first place. Because sonar led the seafloor surveyors to look at this particular underwater structure, there is a certain way in which sound is a property of both the coral ("tinkling") and the technical media of their discovery.[26] Even Hovland's Cold War-Era suspicion was sonically-tuned: that the anomaly might be a "human-made listening device … of the Cold War."[27] Fast-forward nearly 40 years later, and sound remains an important modality for intentionally finding reefs of *L. pertusa*. Synthetic Aperture Sonar—which uses multiple pings to resolve a single spatial location—and multi-beam sonar have been used to scan large areas of the seafloor for *L. pertusa* and create 3-D visualizations, which offer improvements in the two-dimensional images of sidescan sonar.[28] The project AmpLOPHELIA uses passive acoustics to compare healthy and unhealthy reefs of *L. pertusa*, showing how sound can be used as a proxy for gauging reef health.[29]

What is important across these scales of auditory perception is the matter of transduction, or how the technical apprehension of sound is converted into a visual form. In acoustics, transduction is a technical term for the conversation of sound signals from one medium to another. In *Alien Ocean* (2009), anthropologist Stefan Helmreich uses the term transduction to think about a type of ethnographic inquiry motivated by "auditorily inspired, lateral attention to the modulating relations that produce insides and outsides, subjects and objects, sensation and sense data, that produce the very idea of presence itself," a model of "anthropologists as transducers in circuits of social relations" channeling a "mode of attention that asks how subjects, objects, and field emerge in material relations that cannot be known in advance."[30]

Extrapolating from Helmreich, a transductive *literary reading* of Hovland's description is also possible, attentive to how sound signals descriptively move from one medium to another. Consider Hovland's metaphor of *L. pertusa* as glass: "a sigh of astonishment went round as several pieces of white coral bits fell tinkling onto the steel deck, it sounded like bits of glass falling." The first detail, the "sigh of astonishment," is the last sonic effect in this sequence. It would have occurred *after* the coral had resounded, a result of the affective spectacle of listening to and watching bits of coral fall on the ship deck. The sigh among the crew is a kind of *echo* of the tinkling, falling bits of corals themselves, an affective response elicited by the surprise of hearing a sort of deep-sea music. To call the corals glass is retroactively visual, descriptive of *L. pertusa*'s frosted-glass appearance only after the percussive moment of contact with the steel survey vessel. Hovland's passage presents a moment of epistemic *sounding* in the sense of "ascertaining the depth of water by means of the line and lead or (now usually) by means of echo" (OED), both below water (via sonar) and above water (as bits of coral broke and tinkled on the ship deck).

This moment of sounding is inseparable from the infrastructures of petrochemical extraction that led to Hovland's account, situated as it is on the Statoil survey ship, and it contrasts with a more visual relation of coral and glass that Elias identifies in *Coral Empire* (2019). Here, she examines how the photography of John Williamson (in the Bahamas) and Frank Hurley (in Australia) drew upon the optical properties of seawater, a milieu-specific property common in the tropics. Elias suggests that for Williamson and Hurley, the nearly transparent mediation of clear seawater extends the technical function of the glass camera lens: both enable visual perception of corals. We can see an extension of this desire to see *through* the ocean, as if it were air, in contemporary technologies like "fluid lensing," which uses an algorithm to erase the visual distortions of surface ripples for better resolution of the seafloor below.[31] Comparing seawater and camera lenses in this way relies on a notion of environmental mediation via homology—noticing how an environmental feature performs a similar function as anthropogenic technology—a point that John Durham Peters also makes in *The Marvelous Clouds* (2015). Yet Elias notes that glass was important in other ways in underwater photography, which happened:

> through a glass window inside a submersible within an air-filled space; with a watertight camera lowered by poles under the water but operated from above water to avoid the photographer stirring up silt; with a camera encased in a watertight container immersed in water along with the photographer; and through the glass fronts of aquariums.[32]

These many forms of glass—submersible windows, camera lenses, aquarium walls, and even seawater itself—all functioned to mediate the visual apprehension of tropical corals. Here, glass is something one peers through, a technology for focalizing the appearance of bright corals in a milieu that would otherwise blur human vision. In tropical underwater photography, glass both connects and separates the human spectator from the corals, whereas in Hovland's description, *L. pertusa* is (ontologically) the glass—but glass as an acoustic medium, that, when visual, is opaque rather than clear.

This brings me to a key distinction in the media of tropical corals and cold-water corals: tropical corals lend themselves to visual and photographic metaphors in ways that cold-water corals do not. Take another passage in *Coral Empire* where Elias compares tropical corals to cameras: "the natural phenomenon of corals and the technological processes of photographs both required light for photochemical reactions. It was a revelation that corals, as well as photographs, needed light to bring them to life and enable development."[33] Consider this analogy between warm-water corals and cameras. Elias's description of corals as light-sensitive entities characterizes them as a kind of *photomedia*, a capatious term that Erin Despard and Michael Gallagher use to a general expansion of photographic processes beyond the camera—a "strategic abstraction that enables us to identify visual relations in which plants and cameras alike are implicated."[34] I suggest expanding the strategic abstraction of photomedia beyond plants to a broader category of the botanical, such as the symbiotic algae that live in corals, which also have chloroplasts that use sunlight to metabolically produce energy. To elucidate Elias' analogy, then: (tropical) corals are like camera sensors, photosensitive to light because of their botanical symbionts.

We can also find an analogy to visual media in necropolitical descriptions of coral death related to ocean warming. Tropical corals produce fluorescent pigments under moments of stress, which scientists have publicly named "photo-protection." As one article in Nature shows, "in excessive sunlight FPs [fluorescent pigments] are photoprotective; they achieve this by dissipating excess energy at wavelengths of low photosynthetic activity, as well as by reflecting of visible and infrared light by FP-containing chromatophores."[35] The documentary film *Chasing Coral* (2017) described this same vibrant phenomenon as a kind of "sunscreen," and interprets fluorescing corals in striking purple, yellow, and blue hues as "crying for help." Here, the visual analogy shifts from corals being like cameras to corals being like photopaper or film, over-exposed by sunlight in a moment of intensive environmental mediation.

Such media analogies comparing coral to cameras and photopaper do not work for cold-water corals like *L. pertusa*. While photographic technologies (cameras, video, photomosaics) have been used to document *L. pertusa*, Lophelia itself is not camera-like. It does not possess photosynthetic algae at all, and thus lacks the light-sensitivity necessary for a technical analogy to cameras—what Susan Milius playfully calls, in a pun, "corals without boarders."[36] Instead, *L. pertusa* polyps grasp particles of food that float by in the water column, a non-heliotropic metabolism. The symbiotic coral-algae relationship (which *L. pertusa* lacks) is also at the core of how Braverman theorizes coralations, and is symptomatic of a focus on tropical corals.[37] However, Braverman importantly adds that, beyond symbiosis, "coralations also occur at the level of the coral colony and ecosystem, and then at the intersections of culture, science, and law."[38] In this way, Braverman extends a critical consideration of coralations to larger scales of coral relations within a wider ecosystem.

My point is that coralations need to be thought milieu-specifically: the kinds of media analogies that make sense are dependent upon both the environment of the corals and the species. This is important to remember at a time when the default or iconic sense of coral that has dominated most discussions and popularizations is tropical, stony, and symbiotic. Rather, specific corals in specific milieux bring different coralations to the fore—whether they be soft corals, whip corals, hydrocorals, or cold-water corals in a variety of depths unique ecological conditions. Attention to organism and milieu together offers an opportunity to examine assumptions about environmental mediation. *L. pertusa* breaks the correlation (coralation) between corals and visuality: broken glass leads not to an analogy with camera lenses, but with the echoic means of detecting corals through sonar. And sound, as we will see in the next section, is intimately tied to the proximity of oil prospecting in the High North Atlantic.

Coralations of *L. pertusa* and Petroleum Infrastructures

The second coralation I examine is between cold-water coral and infrastructures of petroleum extraction, exemplified in my opening anecdote from Hovland. Indeed, there is a long history of both cooperation and debate between deep sea science and the oil industry, given that the oil industry sometimes volunteers submersible time for the sake of science.[39] Although there has been concern that drilling near *L. pertusa* reefs—which are important nurseries for juvenile fish—could impact future catches, Parmiggiani and Monteiro write that local Norwegian fishermen and oil companies have, at times, both been interested in underwater sensors.[40] *L. pertusa* and other corals are also vulnerable to trawling, a form of fishing that uses weighted nets to scrape the bottom of the seafloor.

One aspect of the coralation between *L. pertusa* and petroleum is geographic: many reefs exist in close proximity to oil seeps (places where oil naturally moves through the seafloor and up to the surface) and subsurface oil reserves, particularly in the High North Atlantic. Yet in addition to a relation of proximity, there have also been cultural choices in how to name particular reefs of *L. pertusa*—including after an oil pipeline. Hovland recounts an experience from 1990, when he was working on mapping a seafloor pipeline route ("Haltenpipe") from the Heidrun oil and gas field, about 200 km off the coast of Norway, to connect with the mainland. During this process, he notes that "many new features were noticed on the seafloor, some of which resembled the Fugløy reef [of *L. pertusa*]" but that it "was not appropriate to interpret them as coral reefs" because they were larger and in greater abundance, so they were simply named "haystacks."[41] However, "There also seemed to be a correlation between abundance of haystacks and suspected seep features, such as shallow 'bright-spots' (strong underground acoustic reflections typical of shallow gas)."[42] Note the instance of transduction, where acoustic reflections are visually described as "bright-spots"—a precondition to later ROV surveillance that did, in fact, confirm that the haystack was a reef of *L. pertusa*. The planned pipeline was rerouted around the reef, and Hovland recalls that, "On the basis of the splendid maps produced by these first ever multibeam echosounder systems, we managed to find a new and improved route."[43] Yet notably, the system of nine live reefs—the oldest of which is 8,150 years old—was *named after the pipeline*, and became Haltenpipe Reef Cluster (HRC).[44] It is telling that Haltenpipe Reef Cluster formally embeds the relations of discovery and proximity between *L. pertusa* and petroleum infrastructures, even as *Halten* also names an old fishing village that is now a tourist attraction.[45]

Some scientists have questioned whether the correlation between seafloor oil seeps and *L. pertusa* also has an aspect of *causation*; in other words, is there something about seafloor sites of oil seepage that is nutritionally advantageous to *L. pertusa*? Hovland argues that there is in his co-written article, "Do Norwegian deep-water coral reefs rely on seeping fluids?" (2003). Here, Hovland and Risk define their "hydraulic theory" as the correlation between *L. pertusa* and the nutrients that are locally available at oil seeps:

> the hydraulic theory [is] for explaining not only the formation and existence of Norwegian deep-water coral reefs, but also their obvious

> wellbeing and proliferation, [which] relies on the assumption that there is a stable, local input of nutrients through the seabed at or near the location where the reefs are found.[46]

Hovland also makes a case for this in *Deep-Water Coral Reefs*, drawing on his experience as a marine geologist finding *L. pertusa* near gas seeps.

Other scientists who study cold-water corals in the Gulf of Mexico disagree, offering a different interpretation: *L. pertusa* may simply like the rocky substrates that are the byproduct of gas-eating microbes, but do not metabolically benefit from oil seeps. Coral biologist Eric Cordes writes that hard corals will still colonize a site where seepage has stopped.[47] Thus rather than substantially relying on oil seeps for nutrients, *L. pertusa* may simply benefit from the byproducts of bacterial metabolisms.[48] Another study by Erin Becker concurs; after examining coral tissues for "tracers of an animal's last meal,"[49] Becker found that coral skeletons did not show evidence of incorporating gas-related isotopes of carbon, nitrogen, and sulfur.[50] The corals were probably not eating any gas products, because the chemical isotopes in their bodies differed from the surrounding rock. These studies suggest that cold-water corals and oil seeps happen to occur in the same place (they correlate), but not because the oil seeps are food for corals (they are not causal).

One final way of thinking about coralations between *L. pertusa* and petroleum infrastructures is through their own agency, given its propensity for actively growing on oil rigs themselves.[51] Because the steel structures in the North Sea have been a "suitable substrata for many fouling species, including L. pertusa," the disused structures have thus been suggested as possible artificial reefs.[52]

That *L. pertusa* grow on oil platforms resonates with how Killian Quigley theorizes encrustation—a process of becoming that "holds together formerly discrete objects and is something more than additive, generating meanings that cannot be accounted for by counting up their parts."[53] Encrustations "attend to the material histories of seabed things,"[54] but also require a methodological approach (encrusting theory) that "assumes that drowned artifactual-ecofactual matter holds and conveys human meaning and memory while supplementing them, and sometimes rearranging them, through temporally, narratively, and ecologically unruly multispecies, animate-inanimate relations."[55] This temporal and spatial aspect of *rearrangement* is what encrustation importantly adds to our milieu-specific thinking about coralations—that two elements (cold-water corals and petroleum infrastructures) not only encounter each other, but also reshape relations.

One aspect of this rearrangement is multispecies in nature: the organisms that coralate with *L. pertusa* differ between cold-water reefs and oil rigs. Bergemark and Jørgenson write that "the soft coral *Alcyonium digitatum* and the anemone *Metridium senile* are two of the most common species observed at oil and gas platforms" but have not been reported from natural *L. pertusa* reefs. Furthermore, the polychaete worm *Eunice norvegica* is absent at offshore oil and gas structures, but is normally associated with *L. pertusa* reefs.[56]

Or perhaps age may play a difference, given that *L. pertusa* reefs can be thousands of years old and form complex multispecies associations, while the communities found at oil platforms are much more recently established.

The coralations between *L. pertusa* and the infrastructures of petroleum extraction thus take a variety of forms—having to do with geographical proximity, with questions of causality, and with the agency of the corals and the multispecies assemblages that they attract. To bring media analogies back into the picture, *L. pertusa* might be seen not only as a keystone species or habitat, but as itself an *infrastructure* for healthy fisheries and benthic communities. "Infra–" after all means "below"—and it is in the undersea depths, sometimes alongside the infrastructures of oil pipelines, that *L. pertusa* silently makes its contributions.

Conclusion

Coralations highlight the lived material realities of ecological assemblages at particular moments in time. In this way they differ from the homophone "correlation," which the Oxford English Dictionary defines as the "mutual relation of two or more things" (in the 16th century) and later in a more statistical sense of correlation as the "interdependence of two or more variable quantities such that a change in the value of one is associated with a change in the value or the expectation of the others" (originating with eugenicist Francis Galton in the 19th century).[57] Media theorist Wendy Chun reminds us that "Correlation's eugenicist history matters … not because it predisposes all uses of correlation towards eugenics, but rather because when correlation works, it does so by making the present and future coincide with a highly curated past," enacting "a future that would repeat their discriminatory abstractions."[58] In thinking about big data, Chun traces the ways in which correlation becomes tied to *prediction*, which can then contribute to reifying categories (such as race, gender, and socioeconomic status). Taking Chun's precautions seriously, we should also consider how the iconic can function predictively, guiding our ontological expectations of corals.

L. pertusa remains the coral that breaks all the rules—not just about where we expect to find corals, but about how they connect to senses of mediation and infrastructures of petroculture. Deviating from the iconic imaginary of tropical corals, *L. pertusa* drifts away from a visual epistemic mode (knowing corals through visual technologies) and a visual ontology (drawing analogies between coral bodies and photographic technologies). Further, *L. pertusa* in the High North Atlantic Ocean draws attention to new coralations with infrastructures of petroleum extraction. Coralations have to be thought in milieu-specific and species-specific ways, attuned to the conditions of mediation in history-saturated oceans.

The iconic imaginary of coral has been useful in cultivating a widespread awareness and care for corals amongst global publics, drawing on their visual beauty as well as functional qualities like ecosystem services. However, in this iconic mode, tropical corals still obscure the kinds of coralations and mediations that arise with the many other corals around the world—standardizing and homogenizing a sense of coral to a single type in a single habitat. When we center other corals like *L. pertusa* as the protagonists of our inquiry, other kinds of relations come to the fore, such as those with petroculture, sound, and transduction. Reexamining the iconic imaginary of corals through attention to particular organisms in particular milieux should lead, then, to a variety of new and unexpected environmental narratives that could also be useful in capturing public attention.

How one might use media and narrative to cultivate public care and concern for *L. pertusa*? One wouldn't use the rainbow iconicity of tropical reefs, nor could one rely on what historian Carolyn Merchant would call a "fall narrative" of Edenic tropical reefs in need of protection. *L. pertusa*'s predominance in deep environments goes against the kind of Edenic images that circulate in the public imaginary.[59] Further, *L. pertusa* is not easily subsumed into the imagination of touristic spaces, existing as it does in deep waters outside easy photographic and oceanic access, and often proximate to oil prospecting and pipelines. Whereas tropical reefs might be visited by swimmers or divers, to encounter more than just a few pieces of *L. pertusa* washed ashore requires advanced technical means and ship time; and to look for *L. pertusa*, one must use technical prostheses (sonar) to listen, and then transduce that acoustic data into a visual map. Like many oceanic organisms and phenomena, making *L. pertusa* tangible to a wider public must negotiate the problem of distance through the affordances of acoustic media and forms of remote observation. Cultivating public care for *L. pertusa* would mean focusing on its unique media: perhaps the ecoacoustics of *L. pertusa* reefs, or macroimages of its tufted polyps, situated in new narratives.

Notes

1 Pantone, "Color of the year 2019," https://www.pantone.com/articles/color-of-the-year/color-of-the-year-2019
2 Ann Elias, *Coral Empire: Underwater Oceans, Colonial Tropics, and Visual Modernity* (Duke University Press, 2019, p. 21).
3 For example, the US National Oceanic and Atmospheric Administration uses "cosmopolitan" to mean "globally distributed" in reference to bamboo corals: https://oceanexplorer.noaa.gov/explorations/04alaska/background/bamboo/bamboo.html.
4 "What are Corals?" National Ocean and Atmospheric Administration, https://oceanservice.noaa.gov/education/tutorial_corals/coral01_intro.html.
5 As Irus Braverman reminds us, associations with coral are also culturally specific, giving the example of corals as the origin of life in the Kumulipo, a long Hawaiian chant about creation. See *Coral whisperers: Scientists on the brink* (University of California Press, 2018, p. 2).
6 There are two major classes of corals: Anthozoa, and Hydrozoa. The corals in Anthozoa are further broken down into two subclasses: Hexacorallina (polyps with six tentacles: hard corals, stony corals, true corals, cup corals, star corals, solitary corals, zooxanthellate and azooxanthellate corals, and more) and Octocorallina (polyps with eight tentacles: soft corals, gorgonians, sea fans, sea whips, sea

feathers, precious corals, sea pens, and more). A diverse range of corals from both of these subclasses can be found in warm waters or cold, deep waters.

7 There is currently a debate in amongst cold-water coral biologists about the nomenclature of this species. In the World Register of Marine Species (WoRMS), molecular studies by Addamo et al. in 2012 and 2016 led to the renaming of the coral as *Desmophyllum pertusum*, rather than *Lophelia pertusa*. However, coral taxomonist Steve Cairns cautioned that this renaming of a major coral would cause confusion, and recommends waiting for more studies. I have chosen to retain the Linnean name, which is still widely recognized and was used in the primary source material I quote from circa 2019. See: https://www.marinespecies.org/aphia.php?p=taxdetails&id=135161.

8 A different version of this chapter appears in Melody Jue, *Coralations* (Minnesota Press, 2025).

9 CORALations, https://www.coralations.org/about_coralations/index.htm

10 Nicholas Mangan, "Core coralations," https://suttongallery.com.au/exhibitions/core-coralations-death-assemblages/.

11 Braverman, "Coralations: Back to the breath," 1.

12 *Lophelia* is not alone: scientists have identified at least 672 species of coral without algal symbionts. Susan Milius, "Corals without Boarders," *Science News* 166:6 (Aug. 7, 2004, 88–89).

13 Roberts, H.H. and Carney, R.S. "Evidence of episodic fluid, gas, and sediment venting on the northern Gulf of Mexico continental slope," *Economic Geology* 92 (1997), quoted in Martin Hovland, *Deep-water coral reefs: Unique biodiversity hotspots* (Springer, 2008, p. ix).

14 Roberts, H.H. and Carney, R.S. "Evidence of episodic fluid, gas, and sediment venting on the northern Gulf of Mexico continental slope," *Economic Geology* 92 (1997), quoted in Martin Hovland, *Deep-water coral reefs: Unique biodiversity hotspots* (Springer, 2008, p. ix)..

15 Thank you to Dr. Kelsey Barnhill, a cold-water coral biologist, for explaining other factors besides depth that drive L. pertusa's localities. See also: Fosså, J.H., Buhl-Mortensen, P., and Furevik, D.M. (2002). The cold-water coral Lophelia pertusa in Norwegian waters: Distribution and fishery impacts. *Hydrobiologia*, 471, 1–12. doi: 10.1023/A:1016504430684.

16 Roberts, J.M., Wheeler, A., Freiwald, A., and Cairns, S.D., *Cold-water corals: The biology of geology of deep-sea coral habitats* (Cambridge University Press, 2009, 20).

17 Allen Coral Atlas, https://allencoralatlas.org/.

18 Reef, Cambridge Dictionary, https://dictionary.cambridge.org/dictionary/english/reef. See also, "Cold-water Coral Reefs: Out of Sight, No Longer Out of Mind" (UNEP World Conservation Monitoring Center), 2004, p. 22.

19 Personal email.

20 Coral books in the arts and humanities that focus on tropical, stony corals include Ann Elias, *Coral Empire* (Duke University Press, 2019); C. Ann Klaus, *Drawing the sea near* (Duke University Press, 2020); Greg Dvorak, *Coral and concrete: Remembering Kwajalein Atoll between Japan, America, and the Marshall Islands* (University of Hawaii, 2018); Irus Braverman, *Coral whisperers* (Duke University Press, 2018); Michele Currie Navakas, *Coral lives: Literature, labor, and the making of America* (Princeton University Press, 2023).

21 Hovland uses the older name "Statoil" in the textbook that I quote from, which was published in 2008.

22 Hovland, *Deep-water coral reefs*, p. 16.

23 Hovland, *Deep-water coral reefs*, p. 16.

24 Hovland, *Deep-water coral reefs*, pp. 17–18.

25 Elena Parmiggiani and Eric Monteiro, "Digitized Coral Reefs," in Janet Vertesi, David Ribes, Carl DiSalvo, Yanni Loukissas, Laura Forlano, Daniela K. Rosner,

Steven J. Jackson and Hanna Rose Shell, eds., *digitalSTS: A field guide for Science & Technology Studies* (Princeton University Press, 2019), 303.

26 Hovland also notes that "glass coral" is a common nickname for *Lophelia* among Norwegian fishermen; Hovland, *Deep-water coral reefs*, p. 8.

27 Hovland, *Deep-water coral reefs*, p. 16.

28 "Synthetic aperture sonar (SAS) differs from regular side-scan sonars by utilizing multiple pings to reconstruct a single spatial location in the output image during post-processing. The main benefits over regular side-scan sonar are improvements in the along-track resolution and a range-independent resolution." Ø. Sture, M. Ludvigsen, M. S. Scheide, and T. Thorsnes, "Recognition of Cold-Water Corals in Synthetic Aperture Sonar Imagery," *IEEE/OES Autonomous Underwater Vehicle Workshop (AUV)*, Porto, Portugal, 2018: 1–6, doi: 10.1109/AUV.2018.8729718.

29 "Healthy and unhealthy soundscapes in cold-water coral reefs," https://laurencedeclippele.com/2023/12/06/healthy-and-unhealthy-soundscapes-in-cold-water-coral-reefs/.

30 Stefan Helmreich, *Alien ocean: Anthropological voyages in microbial seas* (University of California Press, 2009, 230).

31 "Fluid Lensing," https://www.nasa.gov/centers-and-facilities/ames/what-is-fluid-lensing/

32 Elias, *Coral empire*, pp. 58–9.

33 Elias, *Coral empire*, p. 20.

34 Erin Despard and Michael Gallagher, "The media ecologies of plant invasion," *Environmental Humanities* 10:2 (2018), 373.

35 Salih, A., Larkum, A., Cox, G. et al. "Fluorescent pigments in corals are photoprotective," *Nature*, 408 (2000): 850–853. doi:10.1038/35048564.

36 Susan Milius, "Corals without boarders," 88–89. Milius plays on the homophone of "borders" (edges) and "boarders" (tenants), perhaps also a play on the well-known NGO Doctors without Borders. Dr. Kelsey Barnhill also points out that best practices for keeping *L. pertusa* in aquaria are to replicate darkness and use only red lights; see: Orejas et al. 2019, "Cold-water coral in aquaria: Advances and challenges. A focus on the Mediterranean," in *Mediterranean Cold-Water Corals: Past, Present and Future, Coral Reefs of the World 9*, eds. C. Orejas, doi:10.1007/978-3-319-91608-8_38.

37 Braverman, "Coralations: Back to the breath," 1.

38 Braverman, "Coralations: Back to the breath," 1.

39 For example, J. Giles, "Oil-rig staff get into marine biology," *Nature*, 439:899 (2006), doi:10.1038/439899a; "Oil companies help marine biologists to explore new frontiers in deep-sea oceanography," https://phys.org/news/2005-09-oil-companies-marine-biologists-explore.html.

40 Elena Parmiggiani and Eric Monteiro, "Digitized coral reefs" in *digitalSTS: A field guide for Science and Technology Studies*, eds. Vertesi et al. (Princeton University Press, 2019).

41 Hovland, *Deep-water coral reefs*, p. 44.

42 Hovland, *Deep-water coral reefs*, p. 44.

43 Hovland, *Deep-water coral reefs*, p. 45.

44 Hovland, *Deep-water coral reefs*, p. 46.

45 My thanks to the editors of this collection for this detail.

46 Hovland and Risk, "Do Norwegian deep-water coral reefs rely on seeping fluids?" *Marine Geology*, 198 (2003): 84.

47 Erik Cordes, "The ecology of Gulf of Mexico deep-sea hardground communities," https://oceanexplorer.noaa.gov/explorations/06mexico/background/hardgrounds/hardgrounds.html.

48 Cordes notes that "some lab evidence for Lophelia having bacteria living in the mucus on the outside of the coral that can use methane or hydrogen sulfide as an energy source and are capable of transferring this to the coral as a source of food,"

but does not estimate it to be a significant source of nutrients in *Lophelia*'s overall diet (personal email correspondence).

49 "Deep sea corals and methane seeps," https://deepseanews.com/2009/07/seeps-lophelia-carbonate-2/.

50 "Deep sea corals and methane seeps," https://deepseanews.com/2009/07/seeps-lophelia-carbonate-2/.

51 Bergemark and Jørgenson, "*Lophelia pertusa* conservation in the North Sea using obsolete offshore structures as artificial reefs," 278. As Paulina Bergemark and Dolly Jørgenson note, "Roberts (2002) found a total of 133 colonies of Lophelia during a single brief survey of just a small part of each structure" and on additional "13 of 14 oil platforms that were examined in the northern North Sea (Gass & Roberts 2006)."

52 Bergemark and Jørgenson, "*Lophelia pertusa* conservation in the North Sea using obsolete offshore structures as artificial reefs," 277.

53 Killian Quigley, *Reading underwater wreckage: An encrusting ocean* (Bloomsbury, 2023, p. 13).

54 Quigley, *Reading underwater wreckage*, p. 20.

55 Quigley, *Reading underwater wreckage*, p. 15.

56 Kelsey Barnhill writes, "I see this relationship as a bit of a 'chicken and egg' scenario. Eunice helps cement Lophelia branches from separate colonies together, thus helping build and enhance the reef feature. Are there no Eunice on rigs because there are only coral colonies and not reefs, or are there no reefs on rigs because there are no Eunice there to help cement and grow the reef?" Personal correspondence. See also: "Reef-Aggregating behavior by symbiotic eucinid polychaetes from cold-water corals: do worms assemble reefs," *Journal of the Marine Biological Association of the United Kingdom*, 85:4 (August 2005), 813–819. doi:10.1017/S0025315405011756.

57 See also Stephen M. Stigler, "Francis Galton's account of the invention of correlation," *Statistical Science* 1:2 (1989), 73–79.

58 Wendy Chun, *Discriminating data: Correlation, neighborhoods, and the new politics of recognition* (MIT Press, 2022, p. 52).

59 Carolyn Merchant, "Reinventing Eden: Western Culture as a Recovery Narrative," in William Cronon, ed., *Uncommon Ground: Rethinking the Human Place in Nature* (New York: Norton, 1995).

References

Braverman, I. (2018). *Coral whisperers: Scientists on the brink*. University of California Press.

Braverman, I. (2022). Coralations: Back to the breath. *Queensland Review*, 28(2), 1–4.

Chun, W. (2022). *Discriminating data: Correlation, neighborhoods, and the new politics of recognition*. MIT Press.

Despard, E., and Gallagher, M. (2018). The media ecologies of plant invasion. *Environmental Humanities*, 10(2), 370–396. doi:10.1215/22011919-7156793.

Dvorak, D. (2018). *Coral and concrete: Remembering Kwajalein Atoll between Japan, America, and the Marshall Islands*. University of Hawaii Press.

Elias, A. (2019). *Coral empire: Underwater oceans, colonial tropics, and visual modernity*. Duke University Press.

Helmreich, S. (2009). *Alien ocean: Anthropological voyages in microbial seas*. University of California Press.

Hovland, M., and Risk, M. (2003). Do Norwegian deep-water coral reefs rely on seeping fluids? *Marine Geology*, 198, 83–96.

Hovland, M. (2009). *Deep-water coral reefs: Unique biodiversity hotspots.* Springer.

Fosså, J.H., Buhl-Mortensen, P., and Furevik, D.M. (2002). The cold-water coral Lophelia pertusa in Norwegian waters: Distribution and fishery impacts. *Hydrobiologia*, 471, 1–12. doi:10.1023/A:1016504430684.

Jue, M. (2024, forthcoming) *Coralations.* Minnesota Press.

Klaus, C.A. (2020). *Drawing the sea near.* Duke University Press.

Merchant, C. (1995). Reinventing Eden: Western culture as a recovery narrative. In W. Cronon (Ed.), *Uncommon ground: Rethinking the human place in nature.* Norton.

Milius, S. (2004). Corals without boarders. *Science News*, 166(6), 88–89.

Navakas, M. (2023). *Coral lives: Literature, labor, and the making of America*. Princeton University Press.

Parmiggiani, E. and Monteiro, E. (2019). Digitized coral reefs. In J. Vertesi, D. Ribes, C. DiSalvo, Y. Loukissas, L. Forlano, D. Rosner, S. Jackson, H.R. Shell (Eds.), *digitalSTS: A field guide for Science & Technology Studies.* Princeton University Press.

Quigley, K. (2023). *Reading underwater wreckage: An encrusting ocean.* Bloomsbury.

Roberts, H.H. and Carney, R.S. (1997). Evidence of episodic fluid, gas, and sediment venting on the northern Gulf of Mexico continental slope. *Economic Geology*, 92, 863–879. doi:10.2113/gsecongeo.92.7-8.863.

Stigler, S.M. (1989). Francis Galton's account of the invention of correlation. *Statistical Science* 1 (2), 73–79.

Sture, Ø., Ludvigsen, M., Scheide, M.S., and Thorsnes, T. (2018). Recognition of cold-water corals in synthetic aperture sonar imagery. *IEEE/OES Autonomous Underwater Vehicle Workshop (AUV)*, 1–6. doi:10.1109/AUV.2018.8729718.

Index

Note: Locators in *italic* and **bold** refer figures and tables respectively, and locators followed by "n" refer end notes.

Made in the USA
Las Vegas, NV
20 September 2026

cc7e137d-f0c8-41b6-8fc4-2cc2ab94d39aR01